Black Gold

Aboriginal People on the Goldfields of Victoria, 1850-1870

Fred Cahir

Black Gold

Aboriginal People on the Goldfields of Victoria, 1850-1870

Fred Cahir

Australian
National
University

E PRESS

ANU
E PRESS

Published by ANU E Press and Aboriginal History Incorporated
Aboriginal History Monograph 25

This title is also available online at: http://epress.anu.edu.au/

National Library of Australia Cataloguing-in-Publication entry

Author: Cahir, Fred.

Title: Black gold : Aboriginal people on the goldfields of Victoria, 1850-1870 /
                Fred Cahir.

ISBN: 9781921862953 (pbk.) 9781921862960 (eBook)

Series: Aboriginal history monograph ; 25.

Notes: Includes bibliographical references.

Subjects: Gold mines and mining--Victoria--1851-1891.
                Aboriginal Australians--Victoria--History--19th century.

Dewey Number: 994.503

Published with the assistance of University of Ballarat (School of Business), Sovereign Hill Parks
and Museum Association and Parks Victoria

This publication has been supported by the Australian Historical Association

Cover design with assistance from Evie Cahir

Front Cover photo: 'New diggings, Ballarat' by Thomas Ham, 1851. Courtesy State Library of
Victoria

Contents

Preface and acknowledgements

This project began with a humble request from my then Masters supervisor Dr Janice Newton to consider delivering a paper at a conference commemorating the 140th anniversary (1994) of the Eureka Stockade. She wanted me to speak about the role of Aboriginal people in that epoch-making event. At the time I was busy researching about the history of early colonial contact between the Wathawurrung people of the wider Ballarat region and the colonisers who usurped them of their lands. Unaware of the bountiful archival material that existed about Aboriginal peoples' roles on the goldfields of Ballarat in the 1850s, I confess that my initial thought was 'this will be one short conference paper'! I was completely taken aback to discover the vast array of primary source material – newspapers, miner's diaries and other archival sources – that clearly showed 'Aboriginal people had had a dynamic influence on the Ballarat goldfields'. Thoughts of researching and publishing further on this topic had to wait whilst I finished my Masters degree, worked as a teacher on several remote Aboriginal communities in northern Australia and got on with my hectic family life.

It was not until 2001 when I had completed my Masters and began teaching Eco-Tourism at the University of Ballarat that I began to seriously consider filling in the huge gap that existed in history books about the role of Aboriginal people on the goldfields of Victoria. Conversations with Professor Ian Clark at the University of Ballarat and Tim Sullivan from Sovereign Hill followed, and gradually the viability of completing a PhD on the topic of 'Black Gold' emerged as a reality.

I must thank Ian Clark for seeing the potential of the project and Tim Sullivan for agreeing to come on board as Industry Partner in what was to be a successful Australian Research Council grant.

Many people at research institutions were pivotal in my quest for what was sometimes just a rakishly thin sentence in a manuscript or rare book. Tim Hogan at the State Library of Victoria was a marvellous 'urger', the volunteer staff at the Royal Historical Society of Victoria pampered me, the librarians at the University of Ballarat were Trojan-like in their assistance and many members of the Ballarat and the wider district Aboriginal community such as Jaara Elder Uncle Brien Nelson befriended me and applauded my efforts in exploring 'our shared history'. Many people have generously contributed advice, time, information and encouragement to this study and l am indebted to all. Special gratitude goes to: my wife Sandy and my six wonderful children for listening ears, endless cups of tea and forbearance of me during hard times; my PhD supervisors, Professor Ian Clark and Dr Anne Begg-Sunter for their many invaluable insights and their guidance, encouragement, patience and friendship throughout the study; colleagues at the University of Ballarat such as Associate Professor Margaret Zeegers for their positive unfailing support; specialist assistance from Dr Laura Kostanski who was a great listening ear; my extended family who 'put me up and put up with me' on my research sojourns, especially Bernie and Robin and Liam and Carmel. I also thank Sovereign Hill (PhD Industry Partner) for their accessibility and financial support; David Bannear at Parks Victoria and Associate Professor David Goodman for support when it was really needed. Finally, I wish to thank Professor Peter Read for helping to transform my thesis into a book, Aboriginal History Monographs and ANU E Press for agreeing to publish it, and Dr Rani Kerin for being an exemplar 'nudger'.

I give thanks to Jesus Christ for sustaining me.

Fred Cahir
April 2012

Introduction

By the time that gold was officially discovered in Victoria in 1851 the Port Phillip Aboriginal Protectorate (1838-1850) had been disbanded, Aboriginal people had been dispossessed of their land by squatters and sheep, and they were now facing a second invasion – gold seekers from across the globe. When, by the mid 1850s, it became clear that gold was literally strewn across Victoria, the rush to the diggings by a mass of humanity began.

This book dispels four common misconceptions surrounding Aboriginal people on the goldfields of Victoria during the nineteenth century: that most Aboriginal people were attached to sheep stations rather than townships; that those few at mining settlements were on the periphery; that those on the periphery were bewildered spectators; and finally, that Aboriginal experiences on the goldfields were primarily negative. This book reveals that Victorian Aboriginal people demonstrated a great degree of agency, exhibited entrepreneurial spirit and eagerness to participate in gold-mining or related activities and, at times, figured significantly in the gold epoch. Their experiences, like those of non-Indigenous people, were multi-dimensional, from passive presence, active discovery, to shunning the goldfields. There is striking and consistent evidence that Aboriginal people, especially those whose lands were in rich alluvial gold bearing regions, remained in the gold areas, participated in gold mining and interacted with non-Indigenous people in a whole range of hitherto neglected ways, whilst maintaining many of their traditional customs. There is also evidence that Aboriginal people from Tasmania, New South Wales, Queensland and South Australia were present on the Victorian goldfields.

Histories of Aboriginal people and mining

Published in the early 1960s, Geoffrey Blainey's history of Australian mining *The Rush that Never Ended*, is studded with references to Aboriginal people in a number of significant capacities. Yet Blainey neglected to synthesise any broad acknowledgement of their part in the saga of Australian mining. Interestingly, the tendency for writers discussing West Australian, Queensland and Northern Territory goldfields history is to be more inclusive of the Indigenous experience than in Victorian histories where the Aboriginal presence is still predominantly held within frontier violence. Detailed historical studies of specific Aboriginal communities mining and participating in the communal life of goldfields are rare. Indeed, the incorrect attribution of gold discoveries to non-Indigenes, such as the famous 106 pound nugget of gold found by Aboriginal people near the Turon, invariably referred to as simply 'Kerr's Nugget', illustrates how

Aboriginal people have been excised from Australian gold history. Generally, most writers restrict their 'Indigenous participation' lenses to the latter part of the twentieth century and almost exclusively to the northern or arid gold-producing regions of Australia.

An important exception is the collection *Gold: Forgotten Histories and Lost Objects of Australia* (2001) edited by Ian McCalman, Alexander Cook and Andrew Reeves. This work departs from conventional assumptions about gold mining and Aboriginal people, namely that gold created an Aboriginal diaspora as people were forced from their territory. The editors and many of the contributors to this collection argue that while Aboriginal people suffered racial vilification and sustained oppression on the goldfields, this did not prevent their active resistance nor their active engagement with the industry. 'Nowhere do we encounter Indigenes as passive victims of gold', writes McCalman. He cites numerous examples of 'extraordinary sagacity, agile resourcefulness' and the harnessing by shrewd Indigenes of European compulsions. A contributor to this collection, David Goodman, proposes that:

> Until we see the vigorous, masculine, democratic politics of the 1850s gold-rush period, with its insistent calls for the land to be distributed amongst 'the people', as part of the same story as the taking of Aboriginal land and the breaking up of Aboriginal families and communities – until we see, that is, that the 'black armband' history of Australia and the history of democratic progress in Australia tell the same story from different perspectives – we will not have fully acknowledged the conflict of historical understandings which reconciliation aspires to resolve.

The inclusion in this book of many firsthand accounts sourced from miner's diaries, letters, reminiscences, newspapers, paintings and sketches enables the reader to see for themselves how important Aboriginal people were to this crucial period of Victorian and Australian history. The richness of the hitherto neglected primary sources – the abundance of evidence of Aboriginal people's presence on the goldfields – is expressed largely via gold miners' voices which reveal that Victorian Aboriginal people played an active and vital part on the goldfields. Their involvement, as seen through the documentary sources, was diverse and included such roles as: police, gold escorts, guides to new goldfields, bark cutters, prostitutes, trackers, posties, child minders, fur merchants, bushrangers, entertainers and prison guards.

We are fortunate that we have sufficient sources to piece together the details of this forgotten history. The importance of Aboriginal people's participation on the goldfields cannot be overstated. Not only has the traditional story of gold (characterised by a mistaken assumption that the 'Aborigines were swept aside') been shown to be untrue, there is now clear evidence that Aboriginal people were conscious actors and active participants in Australia's economic history, rather than passive spectators, or pawns in another culture's game. The historical record presents a striking picture of Aboriginal people in Victoria demonstrating a desire to relate with and interact with the non-Indigenous miners who had invaded their lands. They learnt the language, speech and manners of the miners and tried to incorporate them into their cultural networks. The records further indicate that Aboriginal people expected to be recompensed for use of their land (via gift-giving and other forms of compensation) and that this was generally not understood by the miners, leading to negative assessments of them by Aboriginal people. Many non-Indigenous perspectives of Aboriginal people and Aboriginal culture, emanating from living in close quarters with each other – especially those revolving around 'begging', 'thieving' and 'nudity' – were also negative. What the records show, however, is that these initially negative appraisals could – and did – change through interaction on the goldfields, enabling miners to 'distinguish the features behind the black mask that had before enveloped them'.

In documenting the history of Aboriginal people's involvement in gold mining in Victoria this book does more than redress an historical error or oversight; it presents a challenging new interpretation of a crucial epoch (and legend making era) in Australian history. After establishing that gold mining occurred on Aboriginal land – a truism that nevertheless bears repeating – this book examines the extent to which Aboriginal people in Victoria came to possess a cultural and economic affinity with gold (ranging from the incorporation of gold into creation stories, seeking gold as independent prospectors, and actively avoiding the social dislocation and environmental degradation that the gold rushes heralded); Aboriginal peoples' perspectives on the work of finding gold and the reasons for their attraction to the goldfields (such as new wealth, new sights, new sounds, and new alliances); and how non-Indigenous people perceived Aboriginal people and their input into the race for gold. Without downplaying the extent of violent conflict that continued to occur between Aboriginal people and the newcomers, without denying the high degree of racial vilification and oppression that Aboriginal people continued to suffer, this book nevertheless documents a significant level of cooperative endeavour that suggests that life on the goldfields may have offered a rare moment of respite from the rigours of colonialism for Aboriginal people.

1. Aboriginal people and mining

It is important to note from the outset that whilst the widespread and bloody inter-racial frontier violence in Victoria had reportedly ceased by 1853, inter-generational violent attitudes had not. In Victoria's post-massacre times many Aboriginal people sought (or were forced) to adapt to the colonial hegemony by adopting conciliatory attitudes towards the colonists in a bid to either remain on their ancestral estates or country which they had come to see as their own, such as mission or government stations. An unidentified Aboriginal (Woiwurrung) man provides much needed insight into these troubled and changing times:

> Bad white men have nearly killed all our men and women ... Before the white people came to our country we were all very happy together; but when they came they gave us grog, and it made us mad. Then we became unhealthy, and began to die off.

> At that time there were a great many men, women and children, but now there are but few of us. But since we began to settle and live in our own houses, we have improved much. We are now happier, and glad to see so many children about us. Some are coming home; they are now tired of the bush.

The gold rushes were the precursor for 'a world turned upside down' not just for the immigrant colonists but for Aboriginal people as well. The gold rushes were not uniform or ordered events that can be categorised easily. Non-Indigenous commentators at the time of the rushes testify to the higgledy-piggledy nature of people's movements. Streams of people from socially and racially diverse backgrounds sojourned from one goldfield or gully to another, the search for the precious metal being the only tangible glue in their communal make-up. The only predisposition to shifting to one location or another (and this was a frequent occurrence) was a more favourable report of gold being found. The focus of non-Indigenous commentary about Aboriginal people became, for a time, almost solely concentrated on perceptions of Aboriginal peoples' responses to the work of finding gold.

It is a truism to say that the gold rushes took place on Aboriginal land, yet it is a truth that is not often articulated. There are many colonial testimonies that auriferous areas were an Aboriginal cultural landscape; that indeed an Indigenous landscape is the fundament that underlies the numerous cultural landscapes laid down after first European settlement in the late 1830s. The evidence for this Indigenous spatial organisation is found in nation or tribal groups with distinct languages, associations with key sites, burial sites found in gold mining areas and conferred place names on the landscape. Auriferous areas did not cease to be Aboriginal cultural landscapes. William Howitt, a visitor to the goldfields, recognised abundant evidence of Aboriginal people's long association with sites near waterways that were also coveted by gold miners. He wrote that he saw 'heaps of wood ashes, partly overgrown with grass, and resembling the barrows of the ancient Britons. They are called native ovens … They contain many wagon-loads of ashes and are found all about this neighbourhood, especially near the creeks'. Burial sites were frequently found in proximity to mine shafts; occasionally bodies were found in the hollow of a tree.

The physical presence of Aboriginal people on the goldfields of Victoria was noted by both writers and artists of the period. John Dunlop, one of the earliest miners on the Ballarat diggings (August 1851), observed that 'there was no sign of any one, only a few huts belonging to the natives'. Miner George Sutherland affirmed Dunlop's account when he conceded that miners had infringed on an Aboriginal landscape:

> This was Poverty Flat, about three quarters of a mile from the spot now occupied by Ballarat; and the hut erected by Dunlop may therefore be considered as the first miner's residence in Ballarat. But, solitary as the place was, they soon found on examination that theirs were not the only habitations erected in this region. Several natives' huts were visible in various places.

Other miners put it more prosaically, such as Emily Skinner who thought 'how many centuries they have in their quiet majesty, or perhaps have looked down on the Aboriginal nation, always fills one with a solemn wonder and brings those words "the everlasting hills" to mind'. Not all encounters of Aboriginal funerary rites were so benign: An *Argus* correspondent on the Omeo diggings (20 January 1857) described non-Indigenous miners reacting to Aboriginal mortuary ceremonies, which they found foreign and harrowing:

> The quiet inhabitants of Flooding Creek have been afflicted with the 'blues' for three whole days the result of the melancholy howlings of a few of the Gippsland blacks and their gins, who are lamenting the loss of a number of their tribe slaughtered by the Omeo tribes, at a place called Tongie. It is painful in the extreme to witness the uncontrollable grief of the poor creatures, more especially the gins, who are seen beating their skulls and tearing their faces with their nails frightfully.

Some non-Indigenous prospectors chanced upon Aboriginal people occupying their traditional camping sites near mining areas. In a compilation of old Ballan (central Victorian town) residents' memoirs are numerous references to 'several tribes [probably Wathawurrung and Djadjawurrung] passing backwards and forwards' and to their 'favourite camping places' such as Doctor's Creek where 'Mr Densley saw several of their corroborees … and also witnessed a tribal fight' whereby two Aboriginal people were mortally injured and interred in their traditional manner.

It is incorrect to maintain that Aboriginal people had no concept of mineral value or that 'for the miners, Aboriginal people were invisible, silent and nameless'. Goldfield writers such as JS Prout observed Aboriginal people (presumably Djadjawurrung) *in situ* at the Mount Alexander goldfield in 1852 but not their active participation in the search for gold. Nor did any goldfields writer note Aboriginal people identifying it as a precious metal that they utilised prior to the onset of the gold rush in 1851, but Prout's assumption that they had no prior knowledge of the metal's existence is not corroborated. Prout recorded in his artwork a scene of mild bewilderment:

> The little group of aborigines at our right, carelessly looking on the busy scene before them, causes one to reflect on the singularity of the circumstance, that, although fond to an extreme of possessing as an ornament any glittering substance, the aborigines as far as we know, have never in their wanderings discovered the precious and most beautiful metal.

While there is no evidence that Aboriginal people attached any great economic or spiritual significance to the heavy yellow metal, it makes sense to assume that they 'must have stumbled over gold nuggets prior to European settlement'. This is supported by anecdotal stories such as that provided by Forster who noted the local clan's immediate prior knowledge of where gold was to be found in great abundance: 'They dug it up amongst the yams on Yam Holes Hill – today a part of Beaufort town'. There are instances of gold nuggets being found associated with old Aboriginal sites, well away from auriferous reefs. The Watchem Nugget from near Maryborough (1904) and the Bunyip nugget from near Bridgewater, east of Bendigo, may both have been carried to their recorded place of discovery by Djadjawurrung people. Notwithstanding, white pipe clay was quarried by Aboriginal people across Victoria prior to colonisation for ceremonial purposes and miners commonly reported the presence of 'splendid nuggets thickly scattered over the white pipe-clay bottom'.

Moreover, much evidence shows Aboriginal people quarrying for crystal, greenstone, sandstone, obsidian, kaolin, ochres and basalt across Victoria. In September 1854 three Aboriginal men assisted William Blandowski, a naturalist, to obtain specimens of a number of fauna including a wombat. Blandowski recorded the traditional method of procurement involved digging shafts to a 'depth of twenty two feet' and a Polish miner on the goldfields of central Victoria, Seweryn Korzelinski, wrote that 'Natives pay respect to talismans which consist of small, very clear crystals often found deep in the diggings'. In the Mt William region William Blandowski observed a 'phonolite (tadijem) quarry, upwards of 100 acres' and a quartzose sandstone quarry at Mt Murchison which 'the natives within a radius of 600 miles [1000km] obtain their supply [of millstones]'. WH Gill considered that the Aboriginal quarry

> is not one large mass of stone, as we know a quarry from which material is blasted, but at Mt William it consists of a whole series of small outcrops of diorite, perhaps in number about 150, within an area of about 25 acres … Nearly all of these outcrops show evidences of having been worked upon, and some of the extensive workings are of very great age … The labour of production was immense.

Snippets of Aboriginal's 'testimonies' testify to their knowledge of minerals on their lands. William Little, wrote of how northern Wathawurrung clans traded gold to shepherds prior to the gold rushes of 1851: 'When erst the shepherds saw the virgin gold A-lying shimmering on fair Nature's breast, And how the ignorant aborigines For trifles gave the precious ore away'.

Records also exist of extensive quarrying and commercial-style transactions for quarried stone being carried out by Victorian Aboriginal people prior to and after British colonisation. WE Stanbridge wrote that Temamet Javolich, a Djadjawurrung clan head, was 'no less than a commercial traveler for the sale of suitable stone for axeheads. His blood relationship with numerous tribes gave him access, and he visited the councils of the tribes arranging barter … his stone quarry was on the Charlotte Plains'. Archaeologist-historian Isabel McBryde has written at length on exchange and trade in south-eastern Australia and has identified several first hand sources which cite the pre-colonialist Aboriginal exchange rate for greenstone hatchet heads, quarried from Mt William.

In newspaper articles, reminiscences, letters and diaries are many inferences to Aboriginal miners on Australian goldfields. For example, some gullies, leads or mines are believed to be named after their Indigenous discoverers or at least attributed to Aboriginal people because of their proximity or some fact connected with them. 'Black Protector's Creek', an obvious reference to the Aboriginal Protectorate Station near the goldfields of Franklinford, in central Victoria is an example. Barry Collett's history of South Gippsland recounts how non-Indigenous gold miners named a creek thus: 'one [unidentified Kurnai] man lived by a creek between the Tarwin and Franklin rivers, lamenting his fate with the refrain "poor-fellow-me", by which the creek is named'. Similarly the Queen Mary Diggings were 'named no doubt after the dusky queen [Djadjawurrung elder] of that area'. According to Fairweather a man named 'Nukong was the headman of the "Ya-itma-thang" during the gold rush days, and it is likely that both Mount Nugong, and the mining township of Nugong were named after him'. Since Aboriginal names were commonly 'subject to corruption by whites' it is possible that the unusual names of some leads and mines may have Indigenous origins. Miner JF Hughes affords us an example where his 'black companions corrected me [about the pronunciation of "Barnawartha"], by saying it was "Barra-na-tha"'.

Employment

Meanwhile the pastoral industry was highly dependent on Aboriginal labour. As Edward Bell, Commissioner of Crown Lands in the Wimmera district observed in 1853, conditions for Aboriginal workers were generally good and a degree of flexibility was built into the pastoral industry to accommodate Aboriginal cultural customs.

> Their usefulness to the white population has been very much increased during the present dearth of labour, produced by the attractions of the Goldfields. There is scarcely a station which the natives are in the habit of frequenting, where they have not been more or less employed ... They appear to be gradually acquiring a knowledge of the value of money, and have been temporarily engaged at rates of wages which in ordinary times, would be considered high for emigrant labour. Their migratory propensities are not, however diminished, and even those who have been longest employed on stations, and appear to have acquired a degree of European civilization in dress and habits of living, are not to be debarred the luxury of occasionally throwing off the restraints of civilized life and visiting their accustomed haunts, and joining in the sports and savage (though generally harmless) warfare of their respective tribes. Very few of them have engaged in the search for Gold.

Aboriginal people mostly continued to have access to foodstuffs through hunting and gathering in the traditional manner which explains 'why they regard with indifference their employment by the settlers' in the Goulburn River district because of the abundance of easily procurable foods from the riverine ecosystem. It also explains why some of these people at least did not seek employment on the goldfields. Bain Attwood, in his study on the Djadjawurrung people of central Victoria, argues there is evidence that in response to the shocking conditions wrought by the proximity of the diggings to their estates, that 'many moved north in the wake of the gold rushes in order to avoid these conditions and to join their kin on the lower Loddon'. Yet the miner John Erskine, writing of Indigenous peoples on the Mudgee goldfields in New South Wales, noted on two occasions their invaluable contribution towards building shelters for the whites and also their seeming lack of willingness to labour at mining. Erskine struggled to explain their aversion, more especially in light of the fact that the people were clearly adept at participating in the monetary system:

> A few Australian blacks had been attracted to the spot and were very useful in assisting the white men to build their bark gunyahs but the labour of digging and washing was not of a nature to suit their habits.

> 1 or 2 Australian blacks were lounging about and were said to have been very useful in assisting the diggers to put up their temporary bark huts. None of them seemed tempted to dig for gold on their own account, although they perfectly understood its value and one readily sold me a "boomerang" for a couple of shillings.

Non-Indigenous miners' poor treatment of Aboriginal people on the goldfields, the availability of traditional food in some areas, and the environmental destruction caused by gold mining (discussed in following chapters), also partly explains Aboriginal apparent disinterest in some gold mining fields.

Statements by miners who emphasised Aboriginal people's lazy and unwilling attitude to work at anything requiring exertion were, by their own words, contradictory, as writers often acknowledged that many had been employed washing sheep, riding horses etc. One miner's explanation for Aboriginal peoples' unwillingness to clean or wash for gold – that it was not in their nature, and that only advanced societies toiled at mining – was exploded a little later in his book:

> At Lawson's Creek 16 miles NNE from Mudgee, a Mr. Bayley had come one day on a party of Australian blacks prospecting on the river on his estate (probably the first who had ever made the attempt) and had encouraged them to proceed.

This observation of industrious Aboriginal people directly involving themselves in gold mining and motivated by personal profit, either in parties independent of non-Indigenous people or as members of 'mixed prospecting parties', is by no means an isolated example. Mossman and Bannister deduce that for some, the 'road shared' was just as much a motive as personal wealth. Other explanations abound. Aboriginal peoples' 'great predilection for white money to spend on rum' was often quoted as their sole reason for occasionally being involved on the goldfields in any capacity. Some gold rush period writers such as Robert Caldwell noted, almost reluctantly, that 'a few of them have tried the diggings' and added 'I am not aware that any of them have ever succeeded as diggers'.

So, given the numerous accounts of Aboriginal participation in gold prospecting, how do we explain their apparent (or at least recorded) disdain for it? Arguably it was the incessant toiling they resented, particularly given that similar sentiments had been voiced in the earlier pastoral period about shepherding. It is also the case that there was less incentive to work for miners and pastoralists in situations where traditional foodstuffs were plentiful. Moreover, digging, let alone mining for gold, required tools and equipment, and indeed a licence which many did not possess, or desire to obtain. The degree of Aboriginal participation in gold rush activities was dependent upon where gold was found, Aboriginal desire to remain on country and, not least, their ability to continue traditional lifestyles in the face of a very sudden and large population increase.

The attraction of mining

Aboriginal people, far from being repelled, were often attracted to the goldfields, motivated by factors such as new wealth, new sights, new sounds and new alliances. At the same time, non-Indigenous people in Victorian gold mining society were captivated by the otherness – the exotica – of experiencing Victorian Aboriginal culture firsthand.

Historian Henry Reynolds, writing of the northern Australian goldfields, suggests that Aboriginal people were attracted to gold mining towns not merely for material goods, or exotica, but out of necessity: 'Many family groups were driven in from the countryside by the violence of the frontier, the difficulty of finding enough to eat in their own country, and because they were literally forced off the land by the squatters and police'. To a very large degree Reynold's summation is most certainly true of the squatting period in Victoria, but not the gold rush. Rarely in the gold rush period were there references to Aboriginal people being explicitly forced onto or off the goldfields due to the violence on the frontier. By 1853 the Guardian of Aborigines, William Thomas, affirmed that inter-cultural frontier violence had all but ceased. Thomas confidently reported: 'We may congratulate ourselves that the weapons of opposition between us and our sable fellows are laid aside … We may safely state that loop holes in huts are no more needed'. Antoine Fauchery, a miner at Ballarat, noted that Victorian Aboriginal people were 'Divided into nomadic tribes [clans] made up of fifteen or twenty individuals, they are seen now in the bush, now in the towns, and still more frequently on the diggings, which they visit by preference'. Others believed that Aboriginal people were attracted to the goldfields for the same reasons as the non-Indigenous miners, that is, to get rich from finding gold and to 'knock it down' at an inn: 'The new [mining] area was situated in the hunting grounds of the Mount Cole tribe of [Djabwurrung] aborigines, who with a view of participating in the prosperity, but more especially in the hope of indulging in cheap liquor, shifted camp to our vicinity'. The attraction of new-found wealth was so great, miner William Craig wrote, that a neighbouring clan who he observed was at enmity with the resident clan, shifted into the locality amongst the gold diggings near Ararat in the 1860s.

> Another [clan] was located some fifteen miles distant, and known as the Mount William clan. By a sort of bush telegraphy the latter soon learned that the Coleites were in clover on the new diggings, and notwithstanding the strange [strained] relations that had existed between the tribes for some years through the abduction of a lubra (woman) from the Williamites, the latter soon put in an appearance.

Sharing country

The sources indicate that Aboriginal alluvial gold seeking was predominantly carried out by small groups, granting moments of some earnings, much excitement and some camaraderie among the miners, black and white. It is difficult to extrapolate whether gold mining offered vestiges of rites of passage initiation, but certainly there were many shared arduous moments and bush mateship that was not at variance with core traditional Aboriginal values. It was often noted in both the mining period and the earlier squatting period how much Aboriginal people enjoyed the thrill, adventure and obvious sense of supremacy and importance they attained when guiding or otherwise sharing their corpus of bush knowledge with whites.

Another factor which explains Aboriginal people's attraction to goldfields was that the existing gold mining areas such as in the Piggoreet area of central Victoria was overlaid on an extant Wathawurrung greenstone quarry, and as such was a highly valued area for the very reasons it was esteemed by the non-Indigenous gold miners. Moreover, the quarry was an important recreational and ceremonial area for the clans who also mined in the area. A local history writer commented:

> For various reasons Piggoreet was a popular camping ground. In the driest of years there was plenty of water and its accompanying animal life. Its prolific vegetation made the marsupials plentiful hence plenty of food for the Blacks. The caves and cliffs gave good shelter from rain and sun. The exposed flints for the making of knives and hatchets must have been a great attraction, as in very few places in Victoria were they so easily exposed as in the exposed cements hereabouts. Most likely Piggoreet, because of its advantages to them, was as popular to the Blacks before the white man's advent as it has been to the said white man, by reason of its romantic scenery and happy days spent in the

height of its mining life. Below Christie's Bridge, though now filled with sand, a very large waterhole, formed by a waterfall over a bar of rock just above it, was a popular camping ground for Blacks.

Some commentators, such as TH Puckle, the Commissioner of Crown Lands based in Hamilton, reported in 1857 that the chief places of resort for the Aboriginal people in his district included the Mt Ararat goldfields, but failed to expand on why the goldfields were frequented by Djabwurrung clans. Likewise, respondents to the 1858 Select Committee on the Condition of Aborigines reported their great attraction to 'frequenting the goldfields'. William Huon, of Wodonga, informed the Committee that in his district, the 'tribes for the last few years have been in the habit of frequenting the various diggings and other townships'. Andrew Porteous, Honorary Correspondent in the Ballarat district, reported in 1866 that 'The Mount Emu tribe still prefer to roam about in small bands, from station to station and the various goldfields'. As late as October 1868 steps were mooted by government officials to induce 'the Aborigines who some time back left Coranderrk [Aboriginal Station formed in 1860-61] and settled on the Alexandra goldfields to return to the station'.

Exotic attraction

The exotic pull of the goldfields and towns for Aboriginal people seems to have centred on the goldfields' social activities: horse races, fetes, galas, official openings, dances, dress, bazaars, unusual animals and new technologies. Charles Fead, a miner on the Buchan diggings, noted at the first local race meeting just a few miles from the diggings the presence of Meteoka, a Kurnai, who had been 'holding horses all day and was proud of it; his honest cheery voice could be heard during the races urging on his favourites'. A Djabwurrung woman, 'Lady Sutherland', was known to frequent the Chute races and it was noted in 1881 that the 'Aboriginal King of Lal Lal was present' at the Lal Lal races. When the train service commenced at Beaufort the *Riponshire Advertiser* reported that 'Jacky Jacky' and his tribe watched the first train go through. Ludwig Becker, diarist of the Burke and Wills expedition in 1860, recorded Aboriginal people in the Terrick Terrick region's first encounter with camels, describing their reaction as one of terror, and of likening the camels to bunyips. John Hunter Kerr, a pastoralist on the goldfields of central Victoria, considered that 'many of the natives were taken at various times to Melbourne and carried to the circus, theatres, or other places of amusement, which must have been as astounding as they were utterly novel to them'. Kerr also witnessed the appropriation by Aboriginal people of 'solitary articles' from miners for their amusement and 'vanity'. Samuel Carter, a squatter in the Wimmera district, recalled taking an Aboriginal companion to the circus in the 1850s. The man was so 'delighted with it that he wished to join the company and I had hard work to get him away', Carter wrote. Charles Fead, a miner in the Gippsland region, noted that Aboriginal people were

> not without vanity and one might occasionally be seen strutting about in a swallow tailed coat or a tall black hat, without another stitch of clothing of any kind. The women too were not a little proud when they could display a parasol and dress improver or, later on, a crinoline for their sole attire. They were fond of looking glasses, bits of finery and scented hair oil.

Aboriginal peoples' fondness for exotic items to 'adorn their persons with' and their great sense of humour and delight in satirising the non-Indigenous peoples' vanity and pompousness was also commented on by John Hunter Kerr, a squatter in central Victoria.

> It used to be no uncommon thing to see some swarthy fellow donning a solitary article of clothing, in comical incongruity with his otherwise perfect nudity. A cravat, a hat, or a discarded crinoline, comprised in some instances the whole of the aboriginal toilet, but was nevertheless sported with great pride and exultation. A gentleman who was subject to frequent attacks of bronchitis one day missed his respirator, without he rarely travelled. After much ineffectual search, it was accidentally discovered in the possession of a black "lubra", who had attached it to her head, and had endeavoured to arrange her dark greasy locks over it in imitation of the "chignons" worn by her white sisters.

An encounter between a (presumably) Wathawurrung man at the Ballarat diggings and a band of wandering musicians highlights the cultural exoticism of the goldfields, as told by Antoine Fauchery, a French miner and photographer:

> It was I think, the first time music was heard on the diggings. An agreeable sensation for all, and particularly novel for the natives. Coloured men, women and children were laughing, foaming, twisting in a general fit of epilepsy. [Only one man] kept his dignity, and neglecting the varied ensemble of the orchestra, all his attention was fixed on the trombone ... it was this mechanism [of the trombone] above all that aroused the lively interest of the observer ... The full extension of the instrument did not over-astonish the black man, but when he saw it drawn back by the instrumentalists hand, go up again, diminish and reduce itself to its simplest proportions he completely lost his head; he touched the brass with his black quivering hands then he came back to the Alsation, on whose person he devoted himself to the most minute researches, opening his coat, thrusting his hands everywhere, but finding nothing.

Suddenly he stopped, enveloped in a fiery gaze the musician and the trombone now all of one piece, then struck his forehead and cried, 'he is swallowing it.' And he ran away, waving his arms in the air, and showing signs of the most dreadful despair.

Others observed Aboriginal responses to non-Aboriginal music: 'They were much interested in Everest's violin and listened to his playing with great pleasure, never having seen or heard such an instrument before. They would approach him with curiosity and examine it carefully, remarking, "Takem box, waddy rub him back, makim noise all same him possum"'. The social etiquette of non-Indigenous people was an exotic experience for some Aboriginal people, one to be savoured as well. Charles Fead, recounted meeting up with 'Metoaka, King of the Omeo Blacks' near the diggings, who with great mirth

> told me, in his own way, of the changes that had lately taken place in his little world – of the erection of a bakery, a restaurant, and a public house, and with a merry laugh, – what l already knew – that a number of white gin immigrants, candidates for domestic service, having arrived at Port Albert, a party of diggers and others had gone down and secured wives a few minutes or, at most, a few hours, after they had met them for the first time in their lives. Such marriages were, at that time not uncommon nor was it a very rare thing to meet with men and women who, living as single, had wives or husbands living, they knew not how or where.

The exoticism of the goldfields cut both ways. Aboriginal people moved quickly to acquire the wonderful contrivances and share in the plentiful goods of the diggings and the townships. Certainly some Aboriginal people suffered intermittent destitution, but an overwhelming body of evidence strongly points to the motive for Aboriginal people soliciting in this period to be one not primarily driven by poverty alone. There were occasional reports of Aboriginal people being hard up for food, clothing and shelter, but most of the evidence points to the fact that Aboriginal people were still largely self sufficient, and when moments of poverty occurred, implored their white brethren for meaningful paid work and keep, rather than simply begging for food and money. A report in the *Grenville Advocate* (2 September 1862) pointed out the unusual occurrence in Linton (Victorian central highlands) of the local Wathawurrung clan who, having a hard winter, gained employment using their traditional skills for a local arboriculturalist.

> The Mount Emu tribe of aboriginals must have been pretty hard pinched for food this winter as they were never before known to be so keen to get employment from Europeans as they have shown themselves this season at Linton. A gentleman of that town … has engaged the tribe to carve him some light-wood uprights for an alcove, as the timber sheds the

bark. It is intended that the carved designs will represent a serpentine coil, similar to that on the shields that the chiefs of the tribe use in times of warfare.

Forming kinship links

The insistent claims on miners such as that described by miner Antoine Fauchery were commonly reported. He wrote of a Wathawurrung man and his three wives who 'skilfully defeated me in a relentless and obstinate battle that went on for not less than two hours'. Fauchery was maddened and confounded by their stubborn begging for some food and wondered 'Had l undergone some magnetic influence?' This frequent occurrence on the goldfields, often construed as begging, was more likely an attempt by clans to obligate non-Indigenous people to rightfully share their possessions with their clan folk. This was a practice that had been utilised by Aboriginal people in the pastoral period in an attempt to assimilate the pastoralists into their social organisation. It is difficult, however, to discern how much of this invoking of kinship ties, as described by miner Walter Bridges at Buninyong (central Victoria), had as much to do with opportunism and how much with the cultural rituals of sharing one's goods.

> My mother and wife and small boy that come out from England with us was standing at the tent one day all alone, no other tents near when they saw a mob of native Blacks and Lubrias [lubras] and a mob of dogs with them come across the gully so my wife said to Mother what ever will we do now so Mother said we must stand our ground and face them for there is no get away So up they come yabbering good day Missie You my countary [country] woman now. My mother had to be spoksman [spokeswoman] the Blacks said You gotum needle missie you gotum thread you Gotum tea you Gotum sugar you Gotum Bacca [tobacco]. So Mother had to say yes to get rid of them and had to give them all they asked for to get rid of them. That was what was called the Bunyong [Buninyong] tribe and when they left they gave their usual salute. Goodbye missie and thankfull enough they was to see them disappear off into the bush.

It is generally agreed that a fundamental response by Aboriginal people in nineteenth century Victoria to the British colonisers was to incorporate them into their kinship networks and thus call to mind their right to resources that were being unjustly denied them. Many correspondents in the colonial period such as Foster Fyans, Police Magistrate at Geelong in the 1840s, had had opportunities to observe closely the strict adherence of Aboriginal people in

Victoria to equality amongst themselves and to ritualised gift giving. Fyans, in a letter to the Colonial Secretary, attempted to explain that an incensed crowd of Wathawurrung clans people besieging Fyan's office was not imploring (begging) the colonial government for food and blankets, they were *insisting* upon it as their right. A number of pastoralists and public servants also reported that they were informed by Aboriginal people that by virtue of their ascribed familial ties, they had moral responsibilities to provide materially for their new 'country men and women'. Aborigines Protector GA Robinson noted when visiting pastoral stations many Aboriginal people were present who recognised non-Indigenous people as their own deceased clanspeople and entered into customary reciprocal arrangements with them.

A number of miners reported relationships forming in spite of their prejudices towards Aboriginal people. Abraham Abrahamsohn, a miner who had set up a 'bakery on a high hill' (Bakery Hill?) in July 1853 on the 'Jurika [Eureka] mines near Pallrad [Ballarat]', wrote: 'The negroid aborigines or Papua, visited me and begged for bread' and to a 'hungry, thin, already elderly Papu, 1 had several times given some of my scanty store of bread and meat and a drink from my bottle of whisky, so necessary in this swamp'. It appears that he was surprised by his own good will adding '1 had even given a chain of glass beads to his young wife. I did this from an unconscious liking for the black man'. Some time later Abrahamsohn had reason to be thankful for his good works towards the 'elderly Papu' as he was visited in an urgent manner by the Aboriginal man who told him

> to expect an attack from one of his own people who was bigger than he, and 1 immediately recalled an unusually big rascal, built only of bone and sinew, whom 1 had caught the day before stealing a knife, and had thoroughly beaten up. I was not exactly comfortable in my isolation, and 1 was overcome with horror to think that when dead, 1 would make a meal for him and his friends.

It is indeed possible that Aboriginal people were emulating the 'hordes' of non-Indigenous beggars euphemistically known as 'sundowners' or 'travellers' who depended on squatters and small land holder's bush hospitality for shelter and sustenance. JH Kerr, a pastoralist on the Loddon River, recalled that 'On my station on the Loddon [circa 1850s], it was no unusual circumstance for twenty, or even thirty of such heterogeneous guests to arrive on one night; while the monthly average rarely fell below 130. Large stations were favoured with greater numbers, all of whom were provided with the staple fare of the Bush – tea, sugar, bread and beef'.

The high number of Aboriginal people during the gold period 'claiming' tribute from non-Indigenous people of rank and position suggests that Aboriginal people still viewed as integral recompense for (and acknowledgement of) their land being usurped, and that 'gentlemen' visitors to the goldfields were an opportunity to seek redress. Many examples exist of Aboriginal people declaring their title to a suite of civil rights such as Equinehup, a Djadjawurrung man, who formally petitioned colonial authorities (Railway Commissioners) expressing his claim to original land title. In 1876 Dicky, a Wathawurrung elder at Lal Lal near Ballarat, complained to some miners that they had 'robbed him of Lal Lal which was his inheritance' and collected several shillings compensation.

Taking advantage

The attractiveness of Aboriginal people to the new immigrant miners centred on their corroborees, weapons, battles, apparel (or lack of), physique, spiritual beliefs, artefacts and athletic prowess. Alexander Finlay, a gold miner on the Bendigo fields in September 1852 marvelled at their ingenuity with a boomerang and skill at ascending trees. At the goldfields near Yackandandah, Aboriginal people were commonly seen exhibiting their boomerang skills to the delight of the miners who paid them cash for these demonstrations. Others were impressed with their ability to 'hit a penny piece 50 yards off' and how their boomerangs moved at 'the speed of lightning, and if aimed true, hitting its victim with an irresistible force'. George Wakefield, a surgeon on the Ballarat diggings, attended 'their corroborees, and their skill in throwing the spear and boomerang is wonderful. I saw the boomerang thrown yesterday, it went completely out of sight and in about 5 minutes returned at the feet of the thrower'.

Striking adornments also punctuated a number of goldfield records: 'Davy saw the other day at the wurlies a black woman ornamented in a manner that l never heard of before. She had kangaroo teeth driven into the flesh above the nails forming a complete set of claws'. C Brout, a Frenchman, was astounded by coats made from platypus skins and necklaces 'made of reeds cut into short pieces, through which threads – also taken from kangaroos' tails – are passed. It is not rare to see necklaces that are eighty to one hundred metres long'. Goldfields newspapers frequently reported on 'native oddities' or merely that a clan's presence in town would occasion a news report of their goings on.

Exoticism was double-edged, that is to say, much fascination and awe was apparent from both non-Indigenous and Aboriginal people about their respective cultures. Walter Bridges at the Buninyong goldfields noted that he

> could not help laughing to see the Blackfellows walking in front like a
> master sweep carrying nothing except a Boomerang or a spear some with
> a old bell topper on and shirt. Then the Luberies [Aboriginal women]

come jabbering along behind carrying the swag in nets some with pups that could not walk others possom skin rugs the Blackfellows make the Lubere do all the work in carrying the loads or baggage.

Certain Indigenous 'notables' became 'familiar figure[s] to the diggers', a reporter from the *Mount Alexander Mail* claimed in 1862. Aboriginal people were often invited (or summoned) to perform for important visitors such as the Duke of Edinburgh, or for commemorative events such as the opening of railways or regattas, to add an air of 'authenticity' and 'novelty'. James Flett suggests that at the opening of the railway at Dunolly 'King Tommy', a Djadjawurrung elder, served as a 'symbol' of indigenisation as he appeared with a banner on a long pole and danced in front of the engine. The Buninyong Council which, in requesting the Governor to insert Buninyong in the programme of places to be visited by the Duke of Edinburgh in 1867, stated:

> That Buninyong is the oldest inland town in the Colony, and the site of the first discovery of gold in Victoria. They have also considered that a corroboree of the Aboriginals would be a novelty to his Royal Highness, and have made arrangements for a large gathering of the Natives for that purpose.

On occasions Aboriginal people were enlisted for non-Indigenous celebrations such as that marking the opening of the Geelong to Melbourne railway in 1853. To mark the occasion a 'procession formed in the Market Square [Geelong] consisting of Mounted Troopers, Police, Soldiers, Railway Navvies, Aborigines, Odd fellows, Laborers, Schools, trades, etc'. Even in the world of theatre stage plays, the presence of Aboriginal people was deemed sufficiently important to display on the bill-boards. 'Off to the Diggings!' a play in London theatres in 1857 featured 'Kikogofatto: A Real Native'.

Not all miners or visitors to the goldfields saw the artistic merit of corroborees. Whilst large crowds of enthusiastic spectators thronged to watch corroborees on the goldfields, some gold seekers such as Emily Skinner were disparaging of such events.

> We came upon a large party of Aborigines at one place, Longwood I think, and they were holding a corroboree, I was told. Certainly they made noise enough. Their dancing and antics were dreadfully grotesque during the short time I watched them. They kept it up till far into the night … They seemed a very miserable degraded people.

Korzelinski, a Polish gold miner opined that it was not in the miner's interest to have Aboriginal people as neighbours at nighttime.

It is not very pleasant to have a blackfellows' camp in close vicinity. On a moonlit night there is no rest for they make a lot of noise and scream in a particular way. Possibly it has a religious significance – a worship of the moon perhaps, or it could be that the *cradje* was giving talismans away. It's hard to say. Tired as laws after a hard day's digging, l was not prepared to flit through the bush at night in the hope of learning some of the natives' habits and ceremonies.

William Craig, a miner at Mount Cole, wanting to be away from the noise and revelry that prevailed at the grog shanty, decided to camp some distance from the diggings only to find that the 'blacks squatted down within a few chains of us, and made night hideous with their barbarous orgies'.

It is evident that the goldfields were places frequented by Aboriginal people and that the attractiveness of the goldfields for Aboriginal people could not be said to be uniform from one goldfield to another. It is equally apparent that Aboriginal culture and lifestyles were an integral part of the goldfields' cultural experience for many miners, evidenced by miners' correspondence, artwork and newspaper reports of the day. These perceptions are interpreted through at times extremely ethnocentric lenses, but demonstrate that for Aboriginal people the goldfields proved to be places which intermittently delighted the pocket and the senses.

2. Discoverers and fossickers

The first discovery of payable gold in Australia has usually been attributed to Edward Hargraves, but there have been consistent reports that others preceded him. John Calvert claimed that with the consistent assistance of his Aboriginal companions he had found gold in New South Wales several years before Hargraves:

> [He got] good results by 'simple crushing and rough washing – with the assistance of his native labourers – Naturally the finder did his best to keep his discovery secret and was for years successful in doing so, having no white allies and treating his black fellows so well as to secure their silence about his searches for the 'medicine earth' … All had gone well so long as he had contented himself with falling back on black labour.

Geoffrey Blainey's history of Australian mining, as mentioned earlier, is studded with references to Aboriginal people in a number of significant capacities. The pivotal roles that Blainey attributes to Aboriginal people included discovering, prospecting and guiding others to some of the most prominent mineral fields in Australia, including the Murchison, the Kimberley, Bathurst, Mount Magnet, White Feather and Tennant Creek. Eyewitness observations from a number of historical sources also reveal that several Victorian diggings were initially discovered by Aboriginal people. Paul Gootch, a non-Indigenous miner at Ballarat, recorded how the rich Eureka Diggings at Ballarat were discovered by an unidentified, presumably Wathawurrung, person. Likewise, Joseph Parker, the son of Assistant Protector of Aborigines Edward Parker, claimed that: 'The first gold in the district [the Loddon valley] was discovered in 1849 by an aboriginal boy in picking up what he supposed to be a stone to throw at a wounded parrot, but it turned out to be a nugget of gold! A European shepherd secured it and kept it secret for two years'.

A plethora of 'how to' books was spawned by the discovery of gold. Writers conveyed to prospective gold diggers the merits and pitfalls of various goldfields and what to take into the bush. Many goldfield promoters discussed the 'Native population' of Australia, and miners such as Charles Ferguson, mining at Linton (south of Ballarat) acknowledged the integral role that Wathawurrung people had played in miners' quests for gold. Miners' accounts record in local histories Indigenous gold miners who struck out successfully on their own. One humorous account revealing the envy displayed towards successful Aboriginal miners, was recorded by Jonathan Moon, who published a short history of Maldon in 1864:

Time and again a member of the tribe would drop in at a local bank to sell a parcel of gold. Knowing ones about town got to hear of this, and considerable manoeuvring went on to win over the confidence of the seller. The blacks maybe were on a "good thing" unknown to all others.

The day came when a certain slick townsman invited Jackie for a ride, and in gleeful anticipation the pair drove off into the country. The merry travellers lubricated at every pub on the way and finally arrived at Newstead. A couple more drinks, then the driver got down to business on Jackie who was thoroughly enjoying himself.

"You a very fine fellow Jackie." Jackie agreed with a wide grin. "You sell em plenty gold?" "Yes, Boss." "Now you tell me where you get the gold and l like you very much." Jackie unabashed and apparently not a bit stupid with liquor, immediately replied: "Boss, blackfellow no b_____ fool!"

Most accounts are of anonymous Indigenous gold seekers. Typical of this record is an account of the diggings in the Evansford district in central Victoria. One digger noted the 'natives learned the value of gold and they soon became searchers for the precious metal'. In 1853 JF Hughes, a digger at Porcupine Flat (near Maldon), wrote that: 'Among those gold-seekers might have been found representatives of nearly every phase of human society [including] the Aboriginal'. Other miners such as William Tomlinson saw Aboriginal people setting off for the diggings from town centres like Geelong: 'I yesterday saw five of the natives in town, they wore blankets over them like shawls. They had all sticks with them. They are perfectly black and very quiet they appear. I saw four of them starting for the diggings this afternoon'.

It was not long after the initial discoveries of Victorian alluvial gold that reports of Aboriginal fossickers occurred with some frequency in both newspaper reports and published accounts of the diggings. There were prolific reports of Aboriginal people forming their own successful gold mining parties across Victoria. The Native Police Corps, who were at first the only police force on the diggings and had prospected for gold at Daisy Hill themselves in 1849, began leaving the force as gold seeking was more attractive. So alluring were the diggings for Aboriginal people that Superintendent of the Native Police Corps, Henry Dana, wrote to Governor La Trobe informing him that he was finding it difficult in May 1852 to prevent the Aboriginal troopers from leaving.

Surprise was often shown that Aboriginal people would show commercial enterprise as it was felt that 'the labour of digging and washing was not of a nature to suit their habits'. F McKenzie Clarke noted that the Aboriginal (Djadjawurrung) fossickers he saw near Bendigo were typical of all miners, in that they sold their gold to purchase subsistence goods such as tea, sugar, tobacco and, flour and beef 'but beef and tobacco were the principal commodities bought'. Clarke further noted the avidity of the Djadjawurrung fossickers who 'during and after every shower of rain, in the gullies, more especially Eaglehawk Gully' were observed.

> There was a camp of blacks [presumably Djadjawurrung people] camped at Myer's Flat [at Bendigo in 1852] they could be seen picking up gold from the red clay and heaps of mullock around the holes. Each one would get a few pennyweights and one called Peter (more civilized than the rest), who had been a bullock driver on a station, could show a few good ounces of gold in a chamois –leather bag, with which he made a trip to Melbourne and imitating the example of a good many of his white brethren, got rid of it in a few days and returned to look for more.

At Castlemaine the *Mount Alexander Mail* (21 March 1862) reported similar observations of a by now very savvy group of Djadjawurrung gold fossickers:

> Aboriginal Fossickers – We noticed the other day a party of native men and women fossicking about the old holes in one of our gullies. Their keenness of sight enables them to detect particles of gold that would escape the observation of most Europeans ... Though too much adverse to steady labour to dig himself for the root of all evil, and probably thinking the white fellow a fool for doing so, the black man does not disdain the yellow dust, when he can procure it for the trouble of picking it up.

Other reports also locate Aboriginal people fossicking for gold over an extended period of time. Some reports emphasised searches of a sustained nature. An inquest held into the death of Fanny Simpson, a Djadjawurrung woman in March 1865 was told 'The Loddon natives had been some time fossicking at Daisy Hill'. A correspondent for *The Argus* reported in 1867 that 'Another instance of good fortune to one of the aboriginals occurred' and further stated that one of the party (Djadjawurrung) had 'stooped down and extracted a nugget, which was sold the same evening for the sum of seven pounds, eighteen shillings and ten pence'. This independent yet intermittent style of gold seeking suggests that there were large degrees of moving in and out of gold mining work at will, a practice that echoed to a large degree their urban work experience and their pastoral work experience. Traveller James Bonwick described an encounter with a newly materially rich Djadjawurrung group which exemplifies and confirms numerous similar accounts of their Indigenous roving and independent lifestyle.

Even the Aborigines are wealthy. I met a party of them at Bullock Ck well clothed, with a good supply of food, new cooking utensils and money in their pockets. One remarked with a becoming expression of dignity "me no poor blackfellow now, me plenty rich blackfellow".

A selection of articles in Victorian central highlands newspapers, reproduced below, spanning eight years (1862-70) clearly demonstrates not just a heightened aptitude for finding gold existed – especially among Djadjawurrung people – but also a commercial attitude. The Djadjawurrung people in central Victoria had had a long and sustained association with Edward Parker, the Aboriginal Protector, and the station at Franklinford (and subsequently were able to live traditional lifestyles and still have the Franklinford reserve as a backstop). This tribe was located in the centre of some of the most spectacular alluvial goldfields and was positioned better than other language groups to engage in mining for gold.

Tarrengower Times, 27 June 1862

A party of aborigines on Wednesday sold a very handsome nugget to Messrs. Warnook Brothers; it was slightly intermixed with quartz, and weighed 7oz. 15dwt, which had been in a hole … The blackfellows were evidently in high glee at their luck.

Argus, 26 July 1864

A party of aborigines had a windfall the other day near Talbot, in the shape of nuggets. Walking over the old ground in Blacksmith Gully, they picked up two nuggets, one weighing a trifle over 1lb, and the other about 1oz. 2dwt. These nuggets had evidently been thrown up from some of the neighbouring claims by the original workers. "Possessed of so much wealth, viz., 51 pounds 14 shillings" says the [Talbot] *Leader*, "the party proceeded to invest themselves in black suits and bell-toppers, and having thus dressed themselves, they swaggered about Amhearst, cutting such airs as to greatly amuse everyone who chanced to see them. The last time they were seen they were trying to make a bargain with Mr Harling for the purchase of a buggy, but the price being beyond their means, taking into consideration the outlay for black suits and bell toppers, they at last requested the loan of a horse and buggy to drive into Talbot, but their wish in this request it appears no one would gratify."

Argus, 6 June 1865

The *Daylesford Mercury* remarks that a small party of aborigines, the remnant, and a very sickly one too, of the Daisy Hill tribe, on Saturday picked up a nugget weighing two ounces on Amherst Flat.

Argus, 7 June 1865

We are indebted to the *Talbot Leader* for the following interesting incident: Our readers will remember the paragraph which appeared in our last issue, notifying that a party of aborigines had found a thirty-ounce nugget at the Emu. This gold realised about 120 pounds for them.

Despite the appearance of luck, it is evident that Aboriginal miners in Victoria were expending effort in a manner that at times closely aligned with Western concepts of time, work discipline, and industrial capitalism. The 'traditional' Aboriginal view of work, that is, maintaining an intense religious relationship with the land and kinship affinities, was not so at odds with the monetary work of alluvial gold mining. Arguably, some groups of Aboriginal miners forged a niche in the dominant economy 'speccing' for gold which did not compromise their ideas of work.

Argus, 21 September 1865

"The Aborigines of this district" says the *Talbot Leader*, "seem to have a peculiar faculty for picking up valuable nuggets of gold. On Thursday, the remnant of the Daisy Hill tribe, while wandering about the old holes in Blacksmiths Gully, Amherst, picked up a nugget weighing six ounces. Mr Douglas of that town, having changed their gold for notes, the party spent about half of the cash upon new clothes, and adjourned with the balance into the bush."

Argus, 1 July 1866

"A party of Aboriginals who have been for the last few days camping near the old lead at Homebush", remarks the *Maryborough Advertiser*, "met with a slice of good luck on Saturday last, in the shape of a four and a half ounce nugget picked up by one of them amongst a heap of tailings in one of the shallow gullies running into the Homebush lead.

Argus, 3 October 1866

"On Saturday morning," says the *Maryborough Advertiser*, "a party of aborigines commenced a search for gold on the heaps of pipeclay at the White Hills, near Mr Mark Drewin's store, and in a very short time they discovered pieces which they sold for 12s., 15s.., and 10 pounds odd. The same party were successful some time since in the neighbourhood of Amherst and Talbot. They say, 'whitefellow dig for gold, and blackfellow pick it up.' Their eyes seem more serviceable than many men's picks and shovels."

As noted earlier a greater number of goldfield correspondents merely noted an Indigenous presence in the vicinity of the gold diggings without any direct attribution to their role as gold seekers. A typical example of miners' references to their presence is CW Babbage's illustration of a 'Blackfellow at the tent, Currency Creek, December 1858' entreating gold diggers to 'Give me sugar'. Newspaper reports too inferred a prominent presence on select goldfields. Buckley, an Aboriginal man (Djadjawurrung) 'and his followers' was reputed to be 'perhaps as familiar to the diggers on the Amhearst and Mia Mia Flat workings as any of the settled residents'. Mrs Charles Clacy, on her visit to the gold diggings of Victoria in 1852-53, emphasised her point, stating that 'All nations, classes and costumes are represented there' including 'Aborigines, with a solitary blanket flung over them'. Later at the Ovens district goldfields William Rayment recorded that 'We fell in with a goodly number of natives at different places on the Road but the most favourite resort of these people are the Public Houses'. Oliver Ragless, a miner at Mount Alexander also had sufficient opportunity to make a number of observations on his neighbours:

> opposite our tent there are a good number of natives [presumably Djadjawurrung] camped. I saw one of them catch an opossum. The men here are generally cleaner than any that I have seen, and rather better looking. They are fond of white clothes and some of them keep them very clean. The women wear their hair long and keep it in good order with a red or blue braid tied around their heads.

The half yearly return of Aboriginal people in the Western Port district for June-December 1852 noted that the two hundred plus Jajowrong [Djadjawurrung] and Malleegoondeet [Barababaraba] tribes 'have been chiefly congregated about the Loddon, Campaspe, and in the vicinity of the diggings at Mount Alexander'. Others such as Morley Roberts emphasised Aboriginal presence in the gold towns, rather than in the bush or on the fields themselves. Roberts wrote a short story about 'King Billy', a Wathawurrung man whose country incorporated Ballarat's goldfields but provides only reference to his urban associations.

> King Billy was given to strolling up and down the streets of Ballarat when that eviscerated city was merely in process of disembowelment, before alluvial mining gave way to quartz crushing … Old Billy was mostly to be found where there was chance of a drink.

Teaming up

The celebrated cosmopolitan nature of the gold diggings has been well documented in the historiography of gold, but rarely is the active Indigenous input alluded to or unpackaged by historians or writers. Yet in the primary records the coming together of 'different races' including Aboriginal people in the search for gold was an often commented upon subject. William McLachlan, in his deposition to the Goldfields Reward Board, acknowledged that when digging for gold in May 1853, a 'blackfellow from Glenorchy' had seen him fossicking at Pleasant Creek (near Stawell) and had then returned with a non-Indigenous miner, 'Dublin Jack', who stated that he intended claiming the reward for finding the first gold. Other miners too remarked on the mixed assemblages of mining parties such as JF Hughes who struck out for the gold diggings in 1853. He exclaimed:

> Porcupine Flat had now rapidly developed into a gigantic rush of some 40,000 people. Among those busy gold-seekers might have been found representatives of nearly every phase of human society – from the Aboriginal, the ticket-of-leave man from the Derwent, the stockman from the Riverina to the enterprising merchant and the Oxford graduate.

The Victorian gold rushes were also responsible for attracting a Tasmanian Aboriginal family to leave Tasmania to settle in the Bangor district in 1853 at Eurambeen Station. John Briggs, his wives Louisa Strugnell Briggs and Ann Briggs, and their family joined the rush to the goldfields, where they lived independently from any government assistance until the early 1870s. In November 1857, John was earning £58 and meals, pitching hay and carting split timber from the mountains. In 1858, his wages rose to £70 and meals. Briggs, not unlike many non-Indigenous workers of the time, went off to the diggings in March but returned to build a new hut on the station, and other bush work such as cutting bark at sixpence a sheet.

William Thomas wrote in 1856 of the large numbers of immigrant Aboriginal people arriving in Victoria during the gold rush period, noting that: 'There have been a greater number of Aboriginals from the neighbouring Colonies than in any other previous year since I have been among them [1838]'. JC Thomson, Crown Commissioner of Lands, alluded to the fact that immigrant Indigenous miners were present on the goldfields of Victoria when he wrote (rather erroneously) that: 'With the exception of one or two brought from distant parts of the country by parties of white men, none have been known to attempt the labour of the goldfields'. Several brief references to immigrant Aboriginal people occur in the historical records, locating them at the goldfields or in close proximity. In 1873, for example, George Brown, an Aboriginal man from Sydney, was arrested near the gold mining town of Steiglitz for deserting his illegitimate child. William Thomas also recorded meeting two gold seekers, one Aboriginal (Thomas Walker) and one non-Indigenous, the Aboriginal man having travelled from the Sydney district with a consignment of horses for Geelong. Thomas Thompson, one of 16 Aboriginal people from Tasmania who joined GA Robinson in Port Phillip, was considered to possibly 'have been on the diggings'. Andy Pittern, an Aboriginal man from Adelaide was known to have lived in Victoria throughout the 1860s but it is not recorded whether he visited the goldfields.

Chinese contributions

It has not been possible to locate any archival records describing Chinese perceptions of Aboriginal people. One fragmentary account of Chinese people by two Queenslander Aboriginal drovers in Victoria is contained within the reminiscences of James Sinclair who wrote that upon the subject of Chinamen being broached, one of them:

> gave us an exhibition of his wonderful powers of mimicry, [he] at once started yabbering away like a Chinaman. So perfect was his imitation of their language that if any person was approaching our camp at the time they would have imagined there was one of the "yellow agony" in our midst. When he finished he laughed heartily and from what he and his father said about the 'longtails' as they termed them, it was evident that they held 'John' in thorough contempt.

The reverend Arthur Polehampton, who spent much time in the Western district of Victoria in the 1850s, considered that 'The blacks are said to have a strong prejudice against the Chinese, whom they accuse of being neither black nor white', and a *Ballarat Star* correspondent reported in 1862 on an 'exchange of insults' between an Aboriginal and a Chinese man in Avoca. Similarly, Peter, a Djabwurrung man, was imprisoned for a week in December 1866 at Ararat for 'assaulting a Chinaman whilst drunk'.

A cartoon which appeared in the *Melbourne Punch* (11 February 1875) titled 'Outraged Majesty' intimates the scorn which Aboriginal people may have held for Chinese people or may simply reflect the racist opinions endemic to large sections of Victoria's non-Indigenous community. A cartoon shows an Aboriginal in a stove pipe hat with nose in the air walking along track and a new chum with swag stopping to ask directions.

> New Chum: "Hi John, is this the right way to Toowambie?"

> King William: "Who you call 'um John, take me for dam Chinaman? Go to the debbil."

A painting by Tommy McCrae, a Waywurru man from north-eastern Victoria, depicting Chinese miners' people being chased by Aboriginal warriors was interpreted by historian Michael Christie as depicting an actual historical event. There is some evidence that Victorian Aboriginal people saw all Chinese people as *'mainmait'* (undesirable foreigners) and therefore McCrae may have painted a scene which took place in an earlier period (1840s) but depicted the attire of Chinese miners in the 1860s. Chinese workers were certainly present in Victoria prior to the gold period, and thus Aboriginal people in some districts were probably at least aware of their presence. In some areas Chinese-Aboriginal work interactions would have been very common, such as at 'Bushy Park' on the Avon River in Gippsland, where in 1854 there was a Chinese cook in the midst of 'a suburban environment of blackfellows' camps'. Illustrations by goldfields artists such as George Rowe clearly depict Aboriginal people in close quarters with Chinese people, demonstrating a large degree of familiarity.

There is also a solitary allusion reported in the *Argus* (21 July 1864) to the possible monetary trade in emu eggs between Aboriginal and Chinese people in Victoria and ample evidence that friendly relations between Chinese miners and Aboriginal people had formed as a result of the attraction to smoking opium by Aboriginal people. The Reverend J Bulmer, under examination by the 1877 Victorian Royal Commission into the welfare of the State's Aboriginal peoples, elaborated on the endemic abuse of opium by Aboriginal people and confirmed they had acquired the habit at the Kiandra diggings, adding that 'they tell me that they buy it from the Chinamen'. According to historian Sue Wesson, the selling of opium by Chinese miners to Aboriginal people occurred in gold mining camps at Yackandandah, Beechworth, Delegate, Craigie, Major's Creek and Nerrigundah.

This pall mall of human society was relatively free of discord: 'You have of course every grade of character amongst the diggers – from the most courteous gentleman to the commonest black – but all seem to harmonize with each other', William Nawton, a miner at the central Victorian goldfields, observed in 1852. An anonymous writer at the Mount Alexander diggings in central Victoria echoed Nawton's observation.

> Where all have much the same aspect and association is necessary for work, while no guarantee of character can be obtained, groupings are formed, not of the most pleasant description to some of the parties. That of a gentleman, two convicts, a black native, and a Zomerzetzire boor, may be taken as a sample.

Unfriendly relations

There were many occasions when relations were anything but harmonious on the goldfields. An anonymous and undated letter written to Philip Johnson recounts a tale of greed and suspicious circumstances on an unspecified goldfield (probably Ovens Valley) of Victoria.

You doubtless remember Black Jim a tall fellow that used to work with Thomson and Jim Lorman he was found dead since you left the blacks are strongly suspected of having murdered him as he was going to the blacks camp when last seen alive he had a good many small nuggets in his possession when last seen amounting in all to about 10oz. So one day the blacks brought in a nugget rather more than an ounce and sold to Mcilveen. So him and a lot of the diggers got the blacks to go out with them to show them where they got it. So the blacks took them up on to the dray road on the side of the mountain and said they got it there in the little ditch made with the dray wheels it was a wet day and there was a stream of water in it. So the people all began to search it. I was coming up from the post office and just as l came to them one of the black fellows pulled his hand out of the water and a nugget in it about 8dwt there was such a rush there as you never saw about twenty or thirty of them all down on their knees shoving one another in the gutters … but no more was to be found. So Thompson saw one of the nuggets and immediately identified it as one the missing man had in his possession when he saw him last so this aroused suspicion and two of the blacks were taken into custody. The commissioner asked them where the man was, one of them said Jacky can tell you. So Jacky another black fellow was immediately seized and ordered to tell. So he took the commissioner right to the spot … the blacks all made their escape the next morning and has never been seen since.

The rationale for having an Indigenous companion on a prospecting party was at times explicitly explained to newcomers by goldfield writers such as James Montagu Smith who believed Aboriginal people were believed by the miners to be more adept at certain tasks than non-Indigenous people. Montagu Smith provides an example:

> We again tried the hole minus two of our hands; and Dick amongst the number showed the white feather, leaving us with only nine good men and true. We repaired the race and set two aborigines at work to cut bark for us, they being so much more ready at it than Europeans.

Writers such as GB Earp wrote effusively of Aboriginal expertise in making huts, keeping the lines of communication open between outstations, locating hapless non-Indigenous stockmen and tracking sheep and cattle. These time-honoured skills were marvelled at and exploited both on and off the goldfields. Short references in miners' diaries relating to the teaming up with Indigenous people are not uncommon and appear to be corroborated by miners' artwork such as 'digger and native'. Thomas Blyth is typical of this occurrence. Having kept a diary on the goldfields of Bendigo and Ballarat in 1852, Blyth devoted two small fragments to his contact with Aboriginal people: 'Proceeded about 3 miles and camped near a gentleman with two blacks … Crossed the Campaspe taking the horses and cart through the R[iver] and paying a native with a canoe to cross our goods'. Other miners were deeply indebted to the Aboriginal assistance proffered to them. A desperate tale of near disaster was retold by George Mackay in which a miner's wife, suffering great privations and on the brink of committing suicide and infanticide in the Loddon River, was rescued by a Djadjawurrung youth.

> According to her own account, she felt impelled to drown herself and her children. Standing there, looking at her shivering, little ones through her scolding tears, and hesitating as to which of them she would throw into the river first, she was startled by a sharp, shrill cry. Turning round, she perceived a young black boy bounding towards her … he quickly explained to her that he was with a dray, which was returning home from an out-station … "No, you cry, Mrs. Charley," said the boy affectionately. "You all right now – directly." "Bless his dear black face," she used to say afterwards in telling her pitiful tale. "It seemed to me like an angel come down from heaven".

George Rowe, at Forest Creek and at Bendigo, also experienced the Djadjawurrung's generosity in the shape of gifts and their willingness to be models for his sketches of them on several occasions. Rowe wrote often of his encounters with the Djadjawurrung, actively seeking them out as he recognised paintings of them were a very saleable commodity.

> I have got some black fellows to draw which I can do by candlelight … A few days since Billy or King William the chief of the Bendigo tribe with one of his gins sat for me for their pictures. I made a sketch of them he had an opossum rug thrown over his shoulders she a blanket … Since I took a sketch of King Billy, I have had a visit from all the tribe every day – they bring me small quantities of gold which they pick up from the surface … they brought me a young kangaroo rat and have promised to get me a young possum … I heard there was some encamped so I walked over and made a sketch of them from which l have today painted a group for 2-2-0 they are a very harmless people … I must try to get hold of some more of the natives and make portraits of them as l find they sell so well … l shut up shop and went over the hills to see an encampment of natives we found them in the forest about sixty to eighty … many of them speak English … This afternoon l walked across the hill into the bush with the intention of taking some sketches of the natives but l found them all gone but 4 old men and a woman with her daughter – of one of the men l took a sketch of them and l hope l shall be able to make money of it.

One of the major setbacks perceived by non-Indigenous people in employing Aboriginal people, reported by employers in pastoral, maritime, service and mining sectors, was the kinship structure that had to be acknowledged by the employer. In the earlier pastoral period squatters and others quickly came to recognise that convivial workplace relations could be often enjoyed only if the families of their valued Aboriginal workers were provided for. JN McLeod of the Western Port district of Victoria, in his reply to the 1858 Select Committee into the condition of Aboriginal people in Victoria, stated that he 'always employed them when he could but found them expensive work people, for if you employed one you had to feed ten'.

Sexual unions

There are few recorded instances of non-Indigenous women openly choosing Aboriginal partners in nineteenth century Victoria. By comparison relationships between non-Indigenous miners and Aboriginal women were not uncommon. Some of these relationships were long-standing ones such as that formed between a Djabwurrung woman from Mount Cole, known as Lady Sutherland, who lived with a miner named Sutherland for 45 years. Dublin Jack, a shepherd and miner on Pleasant Creek in 1853 was reputed to have 'lived six years with the blacks and has two fine sons'. He also had the reputation for being one of the first miners at Pleasant Creek after being advised of the good prospects by an unidentified 'blackfellow from Glenorchy [local sheep station]'. William Thomas, the Guardian of Aborigines reported in 1852 the on-going success (and strains) of an intercultural marriage between a non-Indigenous miner and an Aboriginal woman.

> The white man who many months back was desirous of being married to an Aborigine has kept true to her and she to him, though on account of his occupation, he cannot be continually with her; when he can he returns and brings her clothes and what she requires. She has been in my district from the time she left my roof at Pentridge, and is a kind, faithful, and affectionate servant.

John Morrow relates the story of Annie Griggle, an Aboriginal woman, who came onto the Mitta Mitta diggings, stayed and lived with a non-Indigenous miner (Jack Forrester) and became 'a delight to all her "white piccaninnies" and a respected figure in the valley'.

Many of these relationships between Aboriginal women and non-Indigenous gold miners produced children. One important example is the case of the Connolly family, a prominent Victorian Aboriginal family which is descended from John Connolly. Connolly was born between the years 1855 and 1860 at the Pleasant Creek diggings where his Aboriginal mother (Djabwurrung) lived with a gold digger, and was raised by his maternal great uncle. William Howitt observed that near the banks of the Ovens River

> there was an encampment of natives. Many of them were in the most wretched condition from measles and influenza. We observed that there were whole flocks of children, but nearly all were half-castes. On inquiring how the men tolerated these, we were told, as we had been before, that, up to a certain age, they were very fond of them, and after that they [the children] disappeared!

Wathen also noted the presence of children of mixed Aboriginal and non-Indigenous descent, as did Edward Wilson who wrote in 1859 that: 'These [Murray River] tribes still muster a certain proportion of children amongst them; many of them half-caste.' According to Peter Beveridge, a squatter near present day Swan Hill, by 1861 the only children in the district were those fathered by Europeans. This observation was echoed in other areas and mining regions too. The reportage of mixed descent progeny was often highlighted in government publications. In 1869 the Central Board of Aborigines' sixth report noted that 'at Coranderrk twenty two of the children are blacks, and sixteen are half-castes', though there is little evidence of the Kulin themselves making any such distinction.

The case of 'Buckley', an Aboriginal man murdered on the Mount Alexander goldfields after making petitions to two white miners for two 'lubras' to return from the miners' tents, is in all likelihood an example of non-Indigenous miners abusing the prostitution of, or gift giving, of Aboriginal women that occurred frequently in the earlier squatting period. Seweryn Korzelinski wrote of the ease and affordability of procuring sexual services from Aboriginal women, noting 'a stick of tobacco is sufficient to gain friendship amongst men and favours from women. It is cheap, but then the ladies do not wear lace petticoats'. Robert Young, a commentator on the Victorian goldfields, was told by a teacher at the Aboriginal station at Franklinford in central Victoria that non-Indigenous miners often sought the sexual services of Djadjawurrung women, and that the resultant children were becoming numerous: 'The whites in this locality are continually seeking to seduce the females, and are but too successful; so that the half-caste population is increasing much faster than the natives'.

For Aboriginal women in longer-term relationships, the main benefit was the likelihood of being able to remain in their own country; for the white miner, expert local knowledge and freely-available bush produce. Mossman and Bannister, two miners who wrote about their experiences on the Victorian goldfields in 1853, discussed the constraints and advantages of one such inter-racial union:

Old Bill Cowper would dig away at the side of a mountain and chance it. Very few of the diggers would chance it as Bill did; he never seemed to move from the place where he first commenced. Perhaps it was very inconvenient for him to shift, as he had an Aboriginal woman living with him, which might be a potent reason for his always remaining at one place. Bill had evidently great faith in the mountain... he continued to always dig, digging in the side of the mountain, and washing with the assistance of the Aboriginal woman. A lad who assisted him on one occasion said that he got six hundred pounds worth and that it would keep him and his gin a long time. Bill had been about twelve years beside his hole in the mountain, when l saw him last, and he is likely to die there.

There are some notable examples of fictional liaisons which were published during the gold mining period in which non-Indigenous characters vie for the romantic love of an Aboriginal woman such as Raffaello Carboni's musical play titled '*Gilburnia*'. *Gilburnia* tells how mayhem ensued when miners invaded the traditional territory of the Tarrang tribe and kidnapped Gilburnia, the daughter of the tribal elder. W Dobie also employed this literary sexual entrapment device in his fictional piece 'An Australian Pastoral'.

The relationships that were forged between many non-Indigenous miners and Aboriginal people on the goldfields of Victoria were considered meaningful and educative and it is clear that Aboriginal workers on the goldfields of Victoria were deemed a very desirable adjunct to the non-Indigenous miners who utilised them, but they were generally not an essential condition of success. There were, however, some skilled occupations where Aboriginal traditional knowledge was vital and one of the most notable of these was in the role of guide.

3. Guiding

For the most part, non-Indigenous miners' accounts of directly employing Aboriginal people revolve on the profession of guiding. A guide's role in both the pastoral and gold periods encompassed the most direct and easily traversable route (often along traditional pathways), and locating food, medicine and water to sustain their non-Indigenous companions. John Calvert, 'veteran gold finder of the southern hemisphere', was the recipient of Aboriginal guides' knowledge on many occasions and adamantly avowed that without the aid of his 'blackfellow' guides at Turtle Bay, Western Australia in 1847 'he could never have returned alive'. Indigenous guides also assisted in fording rivers safely, preparing temporary shelters, acting as diplomats and interpreters through the country of resident clans and locating waterholes. There are many instances of Aboriginal people initiating and keenly brokering their work relationships within what could be aptly described as the guiding industry for non-Indigenous gold seekers during the Victorian alluvial gold period (1850-60s). Some guides appear to have taken on the role spontaneously in showing new goldfields, rescuing, providing food, liaising, warning, trading and naming features in the landscape. Although widely praised at the time, they have earned very little recognition by historians.

Historian Les Blake discerned that the early route to the central Victorian goldfields was one blazed by Aboriginals as it had been their traditional pathway.

> The track [to central Victoria] gained clearer definition in October 1851, when a group of intrepid diggers, eager to save time and miles, struck east from Wellington on the Murray. They took aboriginal pads from native well to well and tracks between the few lonely stations in the Long Desert of South Australia to make a fairly straight route to the border. On the Victorian side the Little Desert could be similarly crossed by ways already trodden by both Aboriginals and white men.

Blake's summation is corroborated by miner's accounts such as Oliver Ragless whose party travelled to Mount Alexander from Adelaide only by procuring water from 'native wells'. It is almost certain that in the early period of gold mining, non-Indigenous prospectors were at times following the trading routes/ song-lines of Aboriginal people in the same way that the earlier frontier explorers and squatters had. Franz Micha contends that 'in many cases the modern traffic routes of Whites match the Aborigines' paths of communication, which are thousands of years old'. It is also likely that, as in the pastoral period, in attempting to 'stay on one's country', a number of Aboriginal people attached themselves to groups of miners and at times led them to rich gold bearing sites just as many rich pastoral runs had been opened up initially by Aboriginal guides.

Unwillingness to guide

Sometimes Aboriginal people were unwilling to guide the non-Indigenous travellers. A range of reasons can be discerned. First among these concerned tribal boundaries and protocols. Some guides exhibited a distinct uneasiness when out of their own country. AAC Le Souef wrote of his ('My Nangatta boys') Aboriginal guides 'Tommy and Toby', who were escorting cattle to the Ovens River goldfields, being attacked by 'strange Goulburn blacks'. WT Dawson, the District Surveyor in the Baw Baw and Walhalla region in 1855, discerned in his Aboriginal informants/ guides another reason for their uneasiness. Travelling into certain regions was believed to be life threatening, Dawson explained:

> Although the country is occupied on both sides up to the very base, and in some instances, on a number of the spurs and a portion of the mountains themselves, yet there are peaks and ravines which as yet have never been trodden by a white man. Even the Aborigines recoil with horror-stricken countenances if asked to undertake a journey to the summit of some of them ... Another place which no "blackfellow" will venture is described by them as a boiling chasm which if they approach they get drawn into and are never seen or heard of more.

Similarly, R Brough Smyth recorded how Wathawurrung people reputedly avoided the dark forests west of Mount Blackwood for it was believed that evil spirits dwelt there. An anonymous goldfields' writer in central Victoria explained to colonial readers that Aboriginal people had spiritual and cultural protocols about respecting their dead kinsfolk which meant they could not travel to certain areas.

> Presently I remarked that I had seen no natives about [north of Station Peak, possibly Wathawurrung country]. "No," answered Gower, "their chief, old Warrego died in yonder valley some years ago, and since then no black will spend a night in it. They pass through occasionally, but never stay."

Second, Aboriginal cultural responsibilities took priority over non-Indigenous economic considerations. A party of gold miners who had secured a pact with some Djadjawurrung guides to escort them to Forest Creek gold diggings was disconcerted when their guides informed them that ceremonial rites took precedence over guiding: 'Eager as we were to get away, we were delayed for another evening, in order that a visit from some other friendly tribe might be signalized by a dance. This was their celebrated corroborry'.

The heavily race and class-driven expectations of some non-Indigenous travellers about their Aboriginal guide's role and social standing was also problematic in retaining Aboriginal guides, as work relationships built on a sense of mutual affiliation were commonly considered a critical ingredient for success. Colonial writers such as Francis Lancelott provided advice on the desirability of hiring Aboriginal guides and how to retain them in service. Rules of thumb included:

> Those who have had long experience in the bush are always careful to avail themselves of the services of one or two trusty black attendants … As their services are given more from goodwill than from hope of reward, it is only from attachment to persons with whom they are well acquainted that they are ever prevailed upon to lend themselves as parties in an exploring expedition.

AB Pierce, travelling along the Murray in the early 1860s, forsook this advice and nonchalantly 'hired a black boy for the heavy and dirty work' during the hot and dry season when the 'extreme heat was almost unbearable'. Pierce noted with some indignation (and self-righteous bigotry) that the 'black boy deserted and returned to his tribe, some of whom he met in that neighbourhood – not quite unexpectedly, for desertion is wholly characteristic of the aborigines'.

The perceived usefulness of Aboriginal guides on the goldfields was occasionally limited because they took non-Indigenous people only as far as the borders of their own country. An example of this phenomenon can be seen in Clarke's survey of the Alps in 1858, when three guides left him at or between Muriong and Mowamba and another two left at the junction of the Thredbo and Wallendibby River. Robert Gow gives an account of similar difficulties. Having been left behind for a while, Gow's guide, 'Captain Cadell', confessed to a high level of anxiety in country which he saw as 'too much wild'.

Getting to the goldfields

Edward Snell, a miner trekking from South Australia to Victoria, observed a 'Commissioner under the Pilotage of a Blackfellow going to some newly discovered diggings on the Wimmera'. Sherer, a gold miner in the Omeo district of Victoria also noted a Commissioner of Police and party accompanied by an 'intelligent native' who led them to water, hunted for the party and demonstrated his bush knowledge on numerous occasions during the journey. A tradition of employing Aboriginal guides by Surveyors-General, which had begun with colonisation, was thus continued in the gold mining period to great effect.

It was not solely official surveying parties that were reliant on Aboriginal guides when overlanding to the Victorian goldfields. Members of prospecting parties, such as George Baker, recalled being totally dependent on Aboriginal guides when overlanding from Adelaide to the diggings at Castlemaine in central Victoria.

> We started to go through the "short desert" [probably the Little Desert in north-western Victoria], taking with us two blackfellows with their lubras and picanninies to show us where the water was. It was a very hot and windy day, and we had forgotten to take water with us. It was towards evening when the blackfellows found water, and we were in a very exhausted condition.

John Chapple, on his journey from Adelaide to the Avoca goldfields, also employed a number of Aboriginal guides including 'Black Solomon', 'King Tom' and an unidentified 'black boy', upon whom judging from the repeated references to bush foods, it can be inferred that Chapple's party became reliant.

Reliance on guides

A number of miners were accompanied by Aboriginal guides who on occasion were the actual discoverers of new gold deposits. Historian Barry Collett learned from both oral and archival evidence that small groups of anonymous Kurnai were 'often' members of prospecting parties in South Gippsland, and were at times instrumental in their survival and viability. In 1867, a party made up of three unidentified Kurnai and two non-Indigenous prospectors came close to finding fortune at what was to become the Stockyard Creek diggings. Autumn rains and a lack of supplies forced them to abandon the diggings, reporting they were only able to survive by relying on the meat from koalas and other animals caught by the Kurnai. Fairweather provides a further allusion to the possible influence Aboriginal people played, especially in their role as guides on the more inhospitable goldfields of Victoria, recounting that on one journey to the goldfields 'we escaped Tongio Hill by coming up Swift's Creek (now so called), and had a blackfellow for a guide. Blacks were numerous in Omeo then'.

Despite their importance, Aboriginal guides were often not named, and sometimes not even mentioned. A revealing example of this historical oversight appears in reports relating to the discovery of one of the richest reefs on the Ballarat field. In a letter to the *Geelong Advertiser*, Paul Gootch, a miner in the Canadian and Prince Regent gullies near Ballarat, reported in September 1852 'that the way in which the Eureka diggings were discovered was on the occasion of my sending out a blackfellow to search for a horse who picked up a nugget on the surface. Afterwards I sent out a party to explore who proved that gold was really to be found in abundance'. The lack of recognition given to the unnamed 'blackfellow' was repeated in the case of the Ararat diggings. At the

Linton diggings, south of Ballarat, American miner Charles Ferguson met a large number of the 'Wardy yallock' (Wathawurrung) Aboriginal people in 1851. He reported that:

> There was one black fellow of this tribe who told me he knew where there was plenty of gold, about sixty miles away, and offered to take me or Walter there. We made arrangements to go with him and take one other person also ... They were gone about two weeks. They got gold, but the boys said it was the last place ever made and they would not stop there if they could make a pound weight of gold a day. The same place, but a short time after, turned out to be a good gold district and a great quartz region, known as the Ararat diggings.

Attitudes of the guides and the guided

Some measure of the diversity of interface between Aboriginal people who worked as guides and non-Indigenous miners is portrayed in the following account of a party of miners trekking to new goldfields in the Ovens River area. The extract vividly illustrates the changing relationships and attitudes from both sides of the cultural divide.

> Went to Ovens with Martin Tully, Jimmy Lyons, Andy Clair and Patsy Porker. We had not enough money to buy much tucker. Started on a new track and had no tucker with us. It was raining heavily and we had nothing to eat. Pushed on till we came to a sheep station by chance. Owned by a man named Clarke. He was very good to us. Told us that where we were going there was no tracks. We should take a black guide and got us two from the run on the way. These blacks had a gun each and fearful lot of dogs. We had only one old muzzler loader. Before we went very far they demanded grog. We would not give it, they wanted to walk behind. We were told by Mr. Clarke to keep them in front always. Pestered us for the grog, we gave them some. Then they said they were hungry and set out at a great pace for a tank we could see across the plain. We heard shots and thought that they were shooting at someone. When we got there it was ducks and their dogs were bringing them out. They had a fire lit and cooking the ducks feathers and all, eating them before they were quite cooked. Made Martin Tully sick. We were very anxious to get on and they wanted to camp. One nigger jibbed. King Billy came with us, it was nearly dark, we saw a fire and thinking it might be friends, wanted to go to it but King Billy said "No fear, wild blackfellows kill me all same you" so we cut through the bush. We could not quite feel we could trust him and thought he might be leading us

to these wild blacks. It was very dark and we marveled at him knowing the way he was going. Every now and then someone would ask him "How far?" till at last he said in an impatient voice "Long way yet". About midnight through the thick bush he brought us to a shepherd's hut … Paid old King Billy, who was a good old black. Lyons knocked up and had to go to bed. We went on the next day and old Billy did not notice that Lyons was not with us. He saw his swag on the verandah and thought he had forgotten it and walked three miles after us with it.

Notwithstanding the sometimes difficult cross-cultural negotiations required to recruit Aboriginal guides and the negative attitudes of many miners towards Aboriginal cultural matters, a discernible need was filled by Aboriginal guides thus rendering them indispensable. Sherer acknowledged the considerable effort his party expended in procuring two Djadjawurrung guides: 'For this piece of service we would almost have given all the gold we had'. Cultural misunderstandings were predictable given the preconceived ideas each party often had of each other. HE Haustorfer similarly recounted his great relief at being rescued whilst in the dark bush by unidentified 'blacks' who beckoned him to lie down near their fires. After spending a night under considerable apprehension as he 'felt all sorts of misgivings, thinking they might be longing for a White Roast', in the morning he was relieved when 'the oldest black told his lubra to show me the track'.

The usefulness of Aboriginal guides was not limited to finding a particular place, locating and preparing food, procuring water and carrying supplies. Some miners such as Samuel Lazarus considered the advice of Aboriginal guides to have been of utmost importance to their safe travel in the bush. Bush travellers were often the recipients of Aboriginal bush lore which improved their quality of stay. A story by Katherine McKay about her father indicates how the bush was smoothed for non-Indigenous travellers.

Once Father told us of how, after a long waterless journey, he and his native guide came on a waterhole in an almost dry creek. Father, being very thirsty, took his pannikin to dip a draught out of the film-covered water; but the black guide restrained him warningly, and gathered a bunch of coarse grass that was growing about the creek and placed it on the surface of the water, first dexterously removing a patch of the film or scum, and slowly pressed the pannikin on the filter of dried grass until it was filled with clear water.

Near Bet Bet in central Victoria, Patrick Costello, a shepherd on a pastoral property keen to utilise an Aboriginal guide's expertise and thus save himself unwarranted exertion, asked 'if he would guide me across the hills … To go across country the distance was only about seven miles, whereas if we took the road the distance was 15 miles. The blackfellow agreed to do so, and we started off and reached the station safely'. In the Orbost district, an Aboriginal named Joe Banks rescued a sick non-Indigenous man during the floods by 'making a canoe out of a sheet of bark from the roof and placing the sick man in it, swam through the turbulent waters, towing the canoe and its helpless occupant to safety'.

Water guides

One of the most under-valued contributions Aboriginal people made to the new colonial economy was that of guiding people and stock across the river systems of Australia. Hubert De Castella's description of Aboriginal people guiding large numbers of people, cattle and supplies across the Murray River in the 1850s was a common one: 'Crossing the Murray, which is half a kilometre wide at that spot [junction of the Murray and Darling], was a large number of savages, [who] were camped on the river banks and had boats ready to help the travelers cross'. De Castella described the aplomb and adroitness with which the risky task of guiding people, stock and supplies across was often accomplished.

> The rains had swollen the river so much that Mr. Darchy stayed there camping for a week waiting for a favourable moment to go across. The blacks were particularly useful for transporting men and supplies from the other side of water. They built boats with gum bark, and at spots where roads crossed the river they already had to go a long way to find suitable trees. When the boats were ready the blacks took over a party of men one by one and their horses were sent swimming to them so that they could receive the stock. When the whole herd had crossed the river the men who had stayed behind drove their horses across, and the blacks took them over in turn. It was also the blacks who took across the supplies … Sometimes when the river was not very wide and the current not very swift, one black put himself at the front of the cart and another at the back, and then slipping their heads between the planks it was made of, they would swim across with this heavy load on their backs.

Aboriginal people on the Murray often rescued non-Indigenous peoples' belongings and according to a report (1862) in the *Illustrated Melbourne Post* would consider 'nothing loath to take a good dive, fetching up anything that may [have?] thus found its way to the bed of the river'. Frederick Burchett wrote admiringly of how during floods 'we had to carry rations to outstations in a bark canoe … manufactured by the blacks in a very few minutes'. Many miners and travellers such as Alfred Howitt who conducted geological research in Gippsland (1875) also depended on their Aboriginal guides to construct and pilot vessels for ferrying them across rivers and entrusted them to deliver vital stores and provisions to forward positions.

> I wanted to examine a long portion of the Mitchell River which runs through horizontal strata and which are almost unknown, I therefore sent up two blackfellows "Long Harry" and "Charley Boy" under the care of a trustworthy man to Tabberaberra station at the head of the Gorges. Here they made two bark canoes by the time I arrived from Crooked River and the following morning we started on our voyage ... Long Harry [sat] behind with a piece of green wattle bark in each hand about 6 in. by 12 in. which he used as a paddle ... The other canoe contained Charley and the provisions for three days.

Observers noted the functionality of Aboriginal canoes, not withstanding their less than stately appearance. One reporter noted: 'As a means of crossing the river [Murray] the bark canoe is generally adopted ... although exceedingly primitive, answers every purpose and is often of considerable dimensions, holding four or five, or even six persons'. Many parties were circumspect about the 'primitive constructions' at which they were entrusting their lives but there appears to be no record of them capsizing. One such fear-filled traveller was Edwin Middleton who 'crossed the Murray in a native canoe, a sheet of bark nearly flat. I did not return in it, for l did not relish it, too many black heads bobbing up and down quite close to us. I fully expected [the canoe] to be upset when they caught hold of the canoe, clamouring after tobacco'. AG Pierce, gold miner turned photographer, noted that the

> natives aided us in fording the Serpentine and getting our [photographic] supplies across in their canoes. These boats are of the most primitive construction, being nothing more than a large strip of bark cut to the correct size, with pointed ends, from the eucalyptus tree and dried in the sun, and shaped by a cross stick in each end. The heat of the sun naturally curls the bark and produces a rude boat.

Judging by some accounts, the assistance of Aboriginal guides when fording rivers was not merely time saving. A mother and her child who had slipped into the Campaspe River near Echuca 'would inevitably have drowned, but for an Aborigine known as Tally Ho who happened to be near and who rescued both mother and child'. Similarly, Mrs Campbell, heading to the goldfields, claimed to being saved, along with her sister from drowning whilst crossing a river near Benalla by an Aboriginal guide named 'Captain Cook'. Others such as James Dannock also attested to their indispensability when crossing the Murray.

Dannock, suffering badly from dysentery, and not responding to Aboriginal medicines, entrusted his life to some Aboriginal people who got him across the Murray.

> I took bad with the dysentery and the black lubras [kindly?] got me wattle gum and when l did not get better they said 2 days that fellow go bung [dead] so l thought l had better clear out and got the blacks to put me over the river in a canoe.

Giving directions

Yet another aspect of an Aboriginal guide's job was to track both people and stock, a task which was performed with great regularity and efficiency across Victoria's goldfields. Frequently, passing references (often anonymous) to Aboriginal trackers appear in regional histories. As in the pastoral period, and perhaps more so during the mining era, Aboriginal expertise in guiding lost non-Indigenous people to their desired location was often called upon, and answered. A miner lost in the thick bush surrounding Mount Alexander (central Victoria), recalled how he came upon an encampment of presumably Djadjawurrung Aboriginal people who directed him to Forest Creek:

> I had not gone far before l felt convinced that l had lost my way … Just as it was getting dark, l saw smoke ascending from among the trees … It flashed across my mind that it was the abode of blacks … On getting near, two men of the same description as the former came out, and l inquired the way. They told me l had been walking away from home the whole day. The sun had now set. They gave me the direction of Forest Creek, and away l went into the bush again.

Bush workers in the gold period also relied on Aboriginal people to guide them safely through the bush. Joe Small became lost near the Ovens River and after 'walking some hours' his 'drooping spirits' were lifted by 'a black fellow who gave me full directions respecting my course'. Failed gold miner, GC Fead, also relayed how he was indebted to an unidentified Aboriginal man and three Aboriginal women who guided him and his valuable stock through a region notorious for the risks it posed to human and animal life.

> Down the Jacob's Pinch they [500 head of cattle] went stumbling and sliding much against their will, our horse's feet being almost on a level with their backs. At the foot a blackfellow, with three gins, offered their services to help us through the rocks and we found them very useful.

Once clear of the rocks we camped for the night, the Blacks near us, gladdened with the tobacco and rations which they received as payment for their work.

The surety which Aboriginal guides and trackers afforded non-Indigenous people is vividly illustrated by frequent references in George Sugden's reminiscences of his pioneering experiences both on and off the goldfields in which he relates numerous men and stock being expertly tracked and guided to safety over a period of time.

> "Sugden take Sandy the black boy and see if you can find the man [lost in the bush, and subsequently rescued]". I was quite in his hands and knew that as long as l stuck to him l was safe … a black tracker was employed who can easily pick up your tracks [shepherd rescued] … l was rescued by Sandy the black tracker … rescued by black trackers again … my eyes got so bad that l could not see. I was given a black gin to look after me and lead me about.

Very close personal relationships between Aboriginal guides and non-Indigenous people were established out in the bush. In the letters of drovers and miners there is a palpable bond often born out of being dependent for their lives on their 'sable brethren' in regions where the sourcing of safe drinking water was a matter of life and death. For some non-Indigenous people, as seen in the following account from Robert Gow's journal, the relationship was decidedly a benevolent master-servant relationship, permeated by overtones of 'ownership'.

> Had it not been for my own little black boy – Jacky that would have been the last of me – but he saw the black fellow and gave the alarm to my men. About my Black boy Jacky l will tell you more l have him still he is my right hand man – he saved my life then … he has been with me know [now] 15 years and a more faithful servant no man could have in fact he considers himself as my private property and l can assure you he takes far more interest in my affairs than almost any white man and in many instances he is worth gold to me for he is a splendid tracker … you at home could not credit the way this boy can follow a lost horse or bullock and fetch it … out of the many horses and cattle l have owned l never have lost one since l owned the Black boy.

Assertions of 'ownership' by pastoralists and explorers towards Aboriginal guides were not uncommon, particularly in the pastoral and exploratory periods of colonial Victoria. Representative of this view is Charles Lousada's reminiscences of a selector who

> had a black boy "Toby" with that bush instinct peculiar to the race who he would take away down south towards Lardner and McDonalds track.

Ham [the land selector] would go into the scrub anywhere, and when he had been in a good way, would say to Toby: "Home Toby" and the black boy would bring him straight out.

During the gold mining period, such overbearingly paternalistic assertions of 'ownership' of guides were rare. This is almost certainly because miners – being predominately interested in short-term economic gain – would have viewed such 'possessions' as an encumbrance. It is also likely that the ephemeral state of the goldfields – the lusting after one rich field after another, and consequent crossing of many Aboriginal boundaries – would have acted as a disincentive for Aboriginal people to 'attach' themselves to non-Indigenous brethren. Yet they were needed in the short term. Several testimonies relate the often tragic outcomes of not employing an Aboriginal guide when attempting to traverse from one field to another. Mereweather noted in 1852 'To lose oneself in this district is a serious matter … I hear of many accidents and disasters which have occurred in my district during my short absence in Melbourne'. In such instances, Aboriginal people were frequently employed as trackers to find the missing, as the following chapter demonstrates.

4. Trackers and Native Police

The ability to track or to 'read' the landscape, a highly developed knowledge and skill refined by Aboriginal people, was immediately transferable to the needs of the colonists. Tracking had immediate applications which soon were utilised in many non-Indigenous situations, both before and during (and after) the gold rush period. Gary Presland, in his study *For God's Sake Send the Trackers* evaluated the relationships between members of the Victorian Police and Aboriginal men from Queensland. The Aboriginal trackers:

> exercised skills which were outside the ambit of most Europeans, and the use of which was the major reason for their association with the police. The abilities and knowledge they shared made them a valuable asset to their police employers, and a source of wonder to a wider public. Use of the art of tracking in a context of European law and order is a comparatively recent innovation, but it is only the context which is new; in Aboriginal societies the skills are time-honoured and traditional ones. The expertise displayed by blacktrackers, which has often been described as 'uncanny', 'eerie' and even 'magical', has been developed as an integral part of the complex web of interconnections between people and land which is the fundamental characteristic of Aboriginal society. The use of this craft in a European setting is an example of the way in which Aborigines have successfully adapted the elements of their traditional lifeways to a new world order.

The extraordinary successes of the blacktrackers must have ensured they were loathed and feared by criminals. One of the first instances of the Victorian Police Force calling on the services of Aboriginal trackers from Victoria was during the pursuit of two escaped convicts from Van Diemen's Land in 1853. The two bushrangers had landed in Victoria, committed a series of hold-ups and were successfully tracked to a hold-out in the Gisborne area.

Even after alluvial mining had petered out, the usefulness of blacktrackers proved to be indisputable. Bushrangers such as Harry Powers, one of the most notorious bushrangers in Victoria's colonial history, were traced and captured with the aid of the blacktracker Donald who identified Powers' hide-out when the white officers could not. 'It was then just daylight, and the mist was rolling up the hills, rendering it almost impossible in some places to distinguish it from smoke; but Donald, after one look, pointed straight up the gully, and with dilated eyes and nostrils, uttered in a suppressed tone "[s]Moke! Moke!"'. With the presence of Powers' campfire traced by Donald, Powers was readily apprehended. The murderers of a gold buyer at Omeo in 1860 were likewise tracked and handed over to the police by several unidentified 'natives'. Not just gold thieves were traced by blacktrackers however. On the pastoral stations their tracking expertise was regarded with mystical awe and solemn respect. It was during the gold period that their skills of tracking cattle or sheep were particularly invaluable as a result of the dearth of non-Indigenous workers, due to them deserting to the goldfields. Lawrence Struilby's observations of non-Indigenous people's awed reactions to blacktrackers' 'wonderful power' are typical. It is noteworthy that Struilby recalls that the experienced tracker had his terms of employment brokered by an elder, as per customary law.

> The blacks began to be very useful to us, some of them at least. Some of them had powers of tracking cattle, more surely than a hound would fox or hare; though they did it all by eye. It was amusing to see the chief, after a stiff bargain, hire out a tracker to follow a stray mob of cattle or horses. You would take him to the spot where they were last seen; where he would go deliberately to work to see and measure the track. You must not hurry him … Through scrub and stream, and river and forest, and over sand or rock, he will go, till he brings you to your object, whether it is alive or dead. When he discovers dropping of the cattle, or a blade of grass cut by them, he can tell within a few hours or miles of their whereabouts. Many a fine bullock or heifer they saved for us then; and more for myself afterwards. The kangaroo, opossum, emu, kangaroo-rat, or even the grub, they trace with equal precision. It is as if they could concentrate all their power in the sense of sight.

However, it was the tracing of individuals or parties of people who were lost in the bush which was the most publicly celebrated task which trackers were predominantly called upon to perform throughout the gold period. Francis Lancelott described the pitiful story of a 14 year old girl who had been missing in the bush for ten days before it was 'deemed indispensable' to call for the assistance of black trackers, but it was too late.

> They however, did their part very well. On being told where the girl was last seen to enter the scrub, they went down instantly on their hands and knees, and with their large, sooty eyes, scanned every blade of grass, fallen leaf, and twig, with as much care and delicacy as if they had been objects of infinite worth … it was tedious work for the blacks, but they seemed proud of the great consideration in which their services were held … and, as the blacks had conjectured, her dead body was found on the summit of the rock.

Heart breaking stories such as the one rendered by a Mr Garratt at Little River (near Werribee, south-west of Melbourne) in September 1866 abounded in this period. Garratt reported on the 'assistance of the blacks [Wathawurrung]' being sought by a disconsolate father whose two year old child was lost. Several

Wathawurrung trackers were deployed 'for several days who intelligently and diligently engaged in the search'. Sadly, the child was never found. The most celebrated story of their prowess involved a group of children lost for nine days in the Mallee, successfully tracked by three Victorian Aboriginal men, King Richard, Jerry or Red-cap, and Fred.

It was not just the living whom the blacktrackers were employed to trace but also the dead. The finding of the remains of a lost one brought closure for the parents, family or friends, who otherwise would have cause to 'drink deeply of the cup of sorrow'. Lancelott explained that on the goldfields and elsewhere there was 'always great satisfaction when the remains of the lost are found. Uncertainty is the most calamitous state which the mind can be thrown into. The heart is choked, and there is an unutterable anguish in the pent up and conflicting emotions of hope and fear'. William Thomas noted the valuable service of three Aboriginal people who successfully tracked the body of a murder victim after being called in by the Victorian Police in March 1867:

> You are aware that I was applied to by Mr. Inspector Nicholson of the Detective Force in our [?] On the Subject of Blacks to track Bullarook forest – to find the Body of a Man supposed to have been Murder'd – I furnished Mr. Nicholson with every information – and recommended 3 Blacks [Poker Tommy – Avoca Tribe, Jacky – Ballarat, Billy – Upper Loddon] who were acquainted with that part of Victoria – they succeeded in finding the remains – 20£ was offered reward by Col Secy.

Victorian Aboriginal trackers were commonly called 'police trackers' or 'Native Constables' long after the Victorian Native Police Corps was officially disbanded in 1853. The distinction between tracking work under the guise of the Native Police Corps and tracking work performed on an as need basis was blurred during the gold rush period by writers, and the authorities.

Native Police

One of the major benefits of the Port Phillip (Victorian) Native Police Corps, having ostensibly begun in 1837, was to have at the government's disposal a policing force superbly equipped at tracking criminals in the bush. Some indication of the high esteem that they were held in can be seen in William Strutt's description of them as 'a useful set of men as could be found for special service; particularly tracking in the wild bush carrying dispatches, and they seemed to lend themselves wonderfully to military discipline, and as to their riding and capital seat, you could literally say that man and horse were one'. Beginning in 1849 this role began to shift with the advent of gold discoveries, toward patrolling the new gold finds, guarding the sites, providing order and initially enabling the Port Phillip Government to attempt to keep the gold discoveries a secret. Moreover, it was also in this period that the force began to take part in public celebrations such as the opening of the new Princes Bridge, perform guard duties at Pentridge Gaol and act as official escort to dignitaries. Later members of the corps acted as the first gold escort. They had to ensure the safe passage of large amounts of gold from the goldfields that were both in the possession of private individuals and the government officials who were paid in gold for license fees, a source of revenue for the new Victorian Government.

The Native Police were the first police on the goldfields of Ballarat, (arriving on 20 September 1851) and collected the new goldfield licenses. This new measure (gold licensing) helped to bring in revenue to the new Victorian Government. Stephen Shelmerdine, in his study of the Port Phillip Native Police, considered that by 1851 the Native Police Force was 'operating at its highest level with demands for its services being stimulated by the riot of bushrangers scouring the whole district and the excited fervour of the early goldrush discoveries'. On duty, they accompanied the commissioners on their rounds, and like so much police work their presence alone was important, along with their readiness to intervene in the event of any disorder. Thus it can be seen that the Native Police Corps were briefly at the epicentre of the Victorian gold epoch.

The earliest of their activities was their stint beginning on 5 February 1849 guarding the gold discoveries at Daisy Hill (an outstation located 10 miles west of Deep Creek, one of the branches of the Loddon River). FA Powlett, the Commissioner of Crown Lands for the district, reported that he had left a party of Native Police at Daisy Hill Station to prevent any unauthorised occupation of Crown Lands in the neighbourhood. When the major gold finds at Ballarat, Buninyong, Mount Alexander and Bendigo became public knowledge in 1851, the Native Police, representing the government, were the only effective policing unit Superintendent Charles La Trobe had at his disposal to maintain order and represent the government on the goldfields. William Strutt, a miner on the Ballarat diggings, recalled:

> Met on our way [to Ballarat] a prisoner and a villainous squint-eyed scoundrel he looked, handcuffed and escorted by two well mounted and smart looking black troopers (of whom l have made a drawing), on the road to Melbourne … the useful black troopers were for a time made to escort prisoners to town; these fine fellows were at first the only mounted police; and indeed performed all the police duty at the Ballarat diggings.

Captain Dana, the officer in charge of the Corps, spent three months at the Clunes goldfield, and reported that his troopers picked gold from the ground everywhere they looked. A map depicting the discovery of Bendigo goldfields, holds a small reference to the Corps' presence: 'Black Trooper found spec after 13th Dec [1851]', which demonstrates that they too were caught up in the gold fever. Incidents involving the Corps and miners who resented the licensing fee certainly contributed to their prominence on the central Victorian goldfields. An incident on the Ballarat goldfields on 21 September 1851 illustrates both their success as a force prepared to intervene in cases of disorder and their growing unpopularity in the eyes of miners. Commissioner Doveton and his assistant David Armstrong explained to the diggers the government's decision to introduce licensing fees, which attracted an angry response from the crowd. A public meeting was convened on the spot. The first miners who applied to pay the fee were struck and pelted by 'the mob', as Dana referred to them. Had it not been for the presence of the Native Police, Dana reported, 'those diggers would have been seriously injured'. Following this event a request was made for Native Police then stationed at Goulburn to be redeployed to Ballarat. Cannon argues that the overbearing methods of the Native Police 'so antagonized the diggers that a flame of rebellion was lit, culminating in the Eureka Stockade three years later'. The Native Police Corps Day-Book also demonstrates that in September and October 1852, members of the Native Police were still active, and were accompanying Dana to the diggings at Deep Creek. George Sutherland, a miner at the new goldfield of Ballarat considered them a potent force.

> The Commissioners, Armstrong and Doveton, arrived, and built a small hut on the top of the hill opposite Golden Point. There was also a police officer, named Captain Dana, accompanied by a number of black troopers, ready to support the authority of the commissioners. Going around the ground, they inquisitively looked into each of the claims which were being worked by the industrious diggers.

Many miners noted their presence on the goldfields as conspicuous and adding a touch of exotica to the Ballarat scene such as the following extract from the *Illustrated Australian Magazine* of 1852.

> His [The Commissioner's] tent has the mounted police on one side, and the Native Police, in an extensive mia mia, on the other. The blackfellows are busy tailoring, and here is one on the broad of his back in the sun feigning sleep; and incessantly chattering some monotonous chant ... Close by the Commissioner's tent you observe the encampment of the native police. They too are enjoying the exhilaration of the moment. How graceful are their agile movements. Yonder black fellow is making a feigned attack on his brother with a frying pan; his brother is about to shoot him with his knife. What admirable attributes in both! What dexterous dodging! Frolic is universal among them.

William Brownhill, who found gold at Brown Hill (Ballarat) in 1851, told of how he was caught without a license, taken to the commissioner's camp, and 'guarded by eight or nine black troopers, who in their uniform and polished boots, looked as proud as possible'. One unidentified digger described the 'bustling and picturesque scene' at Mount Alexander in December 1851 when gold to the amount of £25,000 pounds was got ready to be transported to Melbourne.

> The cavalcade consisted of two mounted troopers ahead, then the chaise cart, driven by a officer with an armed guard beside him, and six more troopers on horseback behind, four of them, l think, of the native black police ... on a rising ground the commissioner's establishment is placed, consisting of several tents and two or three gunyahs, or bark huts, made by the native police, after their own fashion. The trooper's horses were standing about ready saddled, and the men themselves, both black and white, and in various costumes, gave life to the picture, while of course some interest was added by the knowledge of the valuable load carried in the cart.

Most miners however were less enamoured by their appearance on the goldfields. George Dunderdale, a miner at Bendigo 'merely glared at them, and let them pass in silence. They were sleek and clean, and we were gaunt as wolves', whilst John Chandler opined that 'They looked enough to frighten anyone; their black faces, big white eyes, long moustache, long swords, carbines, and a pair of pistols in their holsters, was a caution to timid people'. The presence of Aboriginal policemen was condemned by some miners who were already angered with the expensive license system and the overbearing methods rumoured to be used by Dana and some officials including troopers firing upon diggers. Further fuel was added to the hatred towards Dana and the Native Police following an incident reported in the *Argus* in October 1851:

> The redoubtable Captain Dana diversified his exploits on Saturday by knocking down a young man named Thomas with the butt of his whip; the young man fell into a pit from the effect of the blow. It is gratifying to record such a gallant military exploit – a repetition of the like of which will render it a matter of necessity to place him under the surveillance of his own satanic battalion of Black Guards – a suitable troop for such a commander.

Attitudes towards the Corps differed widely. One correspondent lampooned the idea of 'blackfellows' guarding the gold from Ballarat: 'What benefit is it to the diggers to have an escort such as this? One blackfellow leading a horse to which 70 pounds weight of gold is strapped, and two white troopers behind him. A couple of men with double barreled guns might take the gold, blackfellow and horse to boot'. Yet a letter to the *Argus* editor on 26 November 1851 from 'Bucknalook' defended the Corps' efficiency and deplored the crass miserliness of the colonial government towards them.

> A great deal has been said about Christianity and civilizing, this is all talk, talk! Talk of equality of rights! ... The ambiguous captain of this very warlike regiment, it will be seen, figures with 300 pounds [per year] attached to his name ... whilst the efficient part of the company, namely the natives themselves have (Oh! Whisper it not in the same breath with the word justice, mercy, Christianity or equality or rights! THREE PENCE PER DAY!!) Many of these blacks have as correct an idea of the component parts of a shilling, that it is composed of 12 pence as their redoubtable captain, and what must their impression be of this gross act of injustice.

Bucknalook's dire projection proved to be accurate, as by early 1852 Commander Dana was finding it difficult to prevent the Aboriginal troopers from 'absconding', and had trouble attracting new members. 'I can only account for [this]', Dana wrote, 'from the facility they now have of making money, by working for the Settlers, and also from their frequenting the Gold Workings'. An example of this was the sudden desertion of four troopers at Buninyong in October 1851. By leaving behind all their gear and equipment, historian Marie Fels argues, the troopers were clearly signalling their desire to leave the Corps. In December 1851 the Victorian Legislative Council conducted a decisive meeting where the function and future prospects of the Corps were discussed. There were calls for its cessation as it was argued that they were 'utterly useless', whilst others argued it was 'absurd to employ constables whose evidence could not be heard in courts of justice'. However, the continuation of the Corps was secured by the support of both the Colonial Secretary and the Attorney General as reported in the *Argus* (5 December 1851).

> The Committee's erudite attention was directed to the impressive facts that no cases of improper conduct by the Native Police in executing a warrant had ever been reported and that in carrying out normal duties they were as reliable as white men ... In conclusion the Attorney general put forward a thoroughly Australian reason for their continuation – there was no decisive reason for their disbandment at the present stage after so many years in existence.

By February 1852 however Dana had secured the support of Governor La Trobe to radically reform the Native Police Corps. The most important of these reforms was the decision to reduce the number of native troopers, increase the number of non-Indigenous troopers and recruit only native troopers from areas outside Melbourne or Geelong. Dana was also successful in securing La Trobe's support for the native troopers to be used for tracking and escorting rather than policing. It seems clear that Dana believed that the use of Aborigines as troopers had diminished as conflict between Aboriginal people and non-Indigenes had effectually ceased. By October 1852 the Victorian Native Police Corps had finished active duties on the goldfields.

Presland has identified how, on a number of occasions, Victorian Aboriginal people continued to play a role in policing, in matters relating to Aboriginal offenders, bushrangers, tracking lost children (and Victorian Police officers) and tracking lost or stolen horses. In the district of Echuca (1858), an Aboriginal man named Tally-Ho was instrumental in the identification and capture of two Aboriginal men, and again in 1867 three Aboriginal trackers were employed by the Police to find the remains of a murderer. In 1870 an Aboriginal (language group unknown) named Willie Buskin had been working as a tracker in the Kilmore district. At the same time a young Aboriginal man named David was assisting police by tracking the bushranger Harry Powers in the Benalla district. Two instances in the Ballarat district illustrate the use made of Aboriginal (Wathawurrung) people in policing matters. On Christmas Eve 1866, Senior Constable James Mansfield from the Black Hills Station near Ballarat was one of a party of searchers which included three trackers. One of the trackers, named Heath, had been leading the group down a slope when they came upon a recently fallen tree against which a fire had been set. Raking over the remains they recovered some tell-tale pieces of evidence which resulted in a conviction, and subsequent hanging of the accused. In 1867, Jemmy (or Jimmy) Millar (Djabwurrung name: Colit) and Davy Smith, a Tooloora Balug (Buninyong area) clansman of the Wathawurrung tribe were called in to assist police stationed at Rokewood and Buninyong in their enquiries regarding the murder of Thomas Ulrick on the Woady Yaloak goldfields. The two Aboriginal trackers, Millar and Smith, proved to be instrumental in the convictions, leading the police to the guns used in the murder and also tracking the culprit's horse's movements.

It was obvious to some members of the Victorian Police Force that it would be of inestimable benefit to employ trackers on a permanent basis. In March 1859 Henry Hill, an Inspector at Livingstone Creek Police Station in the Omeo district of Gippsland, wrote to his superiors strongly imploring them to consider the overwhelming benefits that the outstanding services of Aboriginal trackers could accrue to isolated police stations on the Gippsland goldfields.

> I have the honour to inform you that on several recent occasions l have seen the urgent necessity of having an intelligent Aboriginal native permanently attached to the police establishment in this district. When in pursuit of the murderers of the late Mr. Green one accompanied us and was mainly instrumental in their capture having tracked them nearly sixty miles to the spot where they were overtaken. At the moment l am

prevented from want of one from searching for the man 'Simpson' lost from Gibbo's Creek. His services would also be frequently made use of for tracing stolen horses etc. For lack of such a guide l am constantly obliged to call upon civilians here to assist us, frequently having to put myself under obligation to persons l should otherwise hold no communication with, and having to divulge my plan of operations where it should be kept secret.

By the 1870s, however, it was believed that most Victorian Aboriginal 'full bloods' were not fit enough, had not retained enough of their tracking skills, and ironically had become too 'civilised' to be useful as trackers for police work. Inspector Hill, in his correspondence also revealed that a lack of trust in local Aboriginal people was probably the principal reason why Queensland Aboriginal trackers, and not their Victorian counterparts, were to be formed into a permanent contingent of the Victorian Police Force. Shellard, a long term resident in the Omeo district maintained that Aboriginal people were members of gangs, that the 'Native camps' were the bases for some bushrangers, and that many Aboriginal people refused to give up information about known identities operating cattle rustling activities in the region.

Consequently, the occasional recruitment of Murris (Queensland Aboriginal people) who had either arrived in Victoria with droving parties or were directly recruited from contacts in Queensland became more frequent. Presland argues convincingly that the hunt for the Kelly gang in 1878 was a turning point in Aboriginal-Victorian Police employment relations that would last until the 1960s:

> There was a significant difference in the way in which the Kelly hunt was handled. In all cases where trackers had been called on prior to 1878, the department had availed itself of local Koori men, who were hired on a short term basis for the purpose in hand. That essentially ad hoc response to investigations was changed forever with the hunt for the Kelly Gang.

It can be seen that, just as the Victorian colonial governments understood the inherent value of employing Aboriginal people as a police force and later as official trackers, so too did non-Indigenous miners clearly perceive the enormous benefits of hiring Aboriginal trackers. The tracking of criminals, lost people and stock was a very highly skilled occupation which, unlike any other, was perceived as the preserve of Aboriginal people, and rarely, if ever, emulated by non-Indigenous people. Hence our historical knowledge of their exploits in this field has been well documented yet, in regional or generalist gold histories, they are rarely accorded the significance that was bestowed upon them during the gold rush period, and remain, to a large degree, outside special exhibitions or publications, as invisible actors on the gold stage.

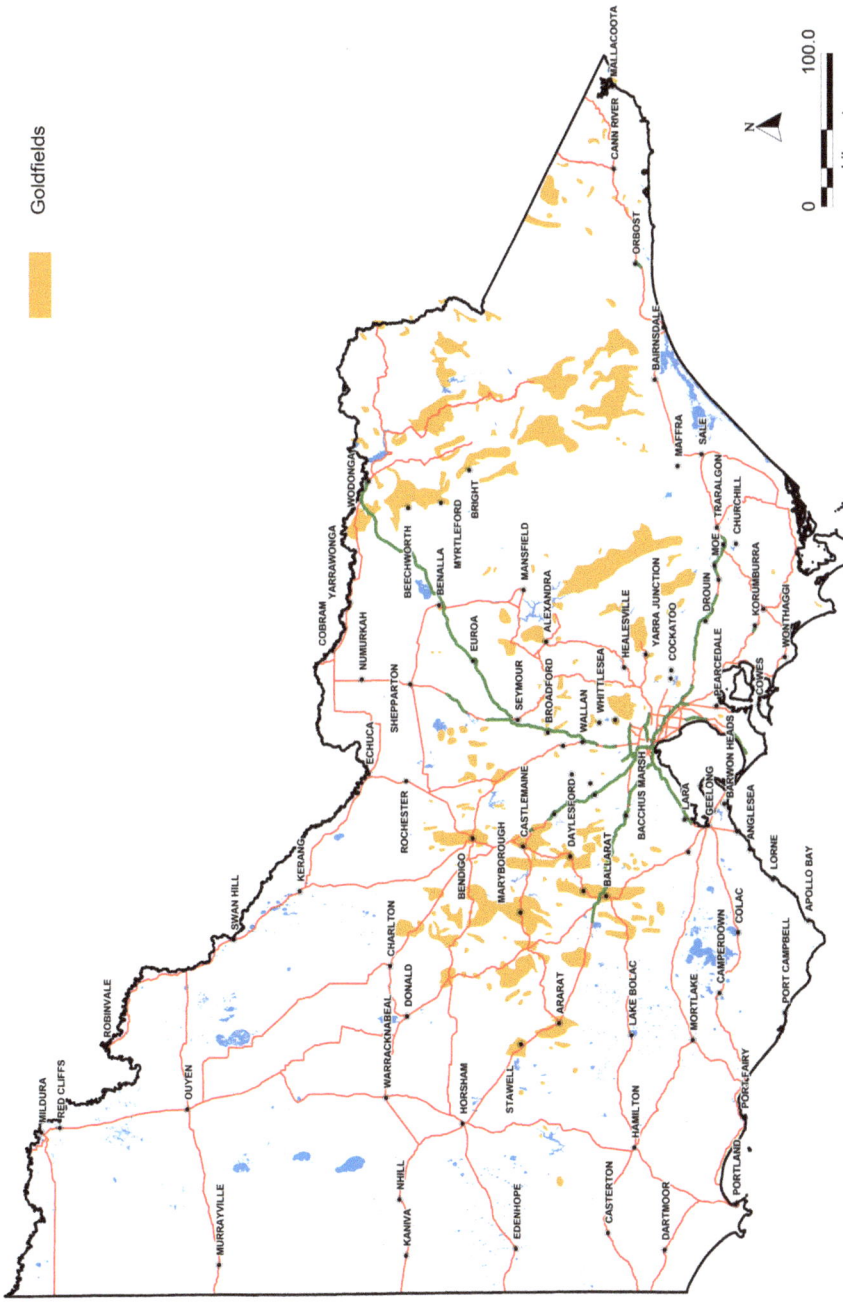

Map of Victoria showing location of goldfields.

Map adapted from Spatial Vision (2001).

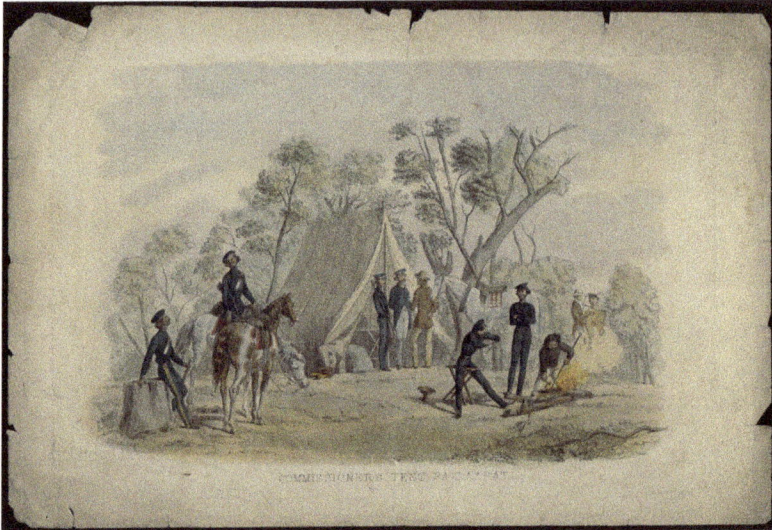

The 'Commissioner's tent, Ballaarat' clearly depicts the central role the native police corps had on the Victorian goldfields.

'Commissioner's tent, Ballaarat', Ham, Thomas, David Tulloch, William Strutt, Cyrus Mason and G Strafford, (1854), *The gold diggers portfolio consisting of a series of sketches of the Victoria Gold Fields*, engraving, Melbourne: T Ham. Rex Nan Kivell Collection, NK11266/F, National Library of Australia, vn3078574.

BUSH MAILMAN.

'Bush mailman' or 'bush postman' shows the critical role Aboriginal people performed in frontier communications during the gold period.

'Bush mailman', Gill, Samuel Thomas, Hamel and Ferguson, (1864), chromolithograph, Melbourne. National Library of Australia, an7149190.

Edwin Stocqueler like many goldfields artists 'travelled widely' in the 1850s depicting Victorian Aboriginal ceremonial life during the gold period.

'Night corroboree of Australian natives', Stocqueler, Edwin, (185-?), oil on canvas. Rex Nan Kivell Collection, NK6777, National Library of Australia, an2282355.

Non-Indigenous miners frequently commented upon the regular occurrence of corroborees being performed during the gold rush period in Victoria. Note the non-Indigenous spectators in the foreground who may have been summoned to view this corroboree.

'A native corroboree at night', Gill, Samuel Thomas, (ca 1850), watercolour. Rex Nan Kivell Collection, NK2124, National Library of Australia, an3366343.

Maintaining law and order on the Victorian goldfields was largely achieved by policing the illegal sale of alcohol. Note the mounted Aboriginal trooper[s] in the background to the right.

'Sly grog selling at the "diggins" effectually stopped', Gill, Samuel Thomas, (ca 1853), pen and ink drawing. Rex Nan Kivell Collection, NK6869, National Library of Australia, an6617940.

Two Wathawurrung people, one of them heavily laden with goods walk along what is now Main Road Ballarat, not far from Golden Point, presumably not long after the 1854 Eureka rebellion. The Red Coat soldier's camp can be seen in the background.

'Ballarat, Victoria, ca. 1854', Gill, Samuel Thomas, (1850), watercolour. Rex Nan Kivell Collection, NK166, National Library of Australia, an6618003.

Djadjawurrung/Jaara people were a very visible presence on the goldfields of Mt Alexander. Newspaper reports and miner's letters attest to their active participation in the rush for gold.

'Mount Alexander from Saw-pit Gully', 'An old colonist', (1856), National Library of Australia, an9092003.

Goldfields' artist George Rowe noted that he had many interactions with the Aboriginal people (Jaara/Djadjawurrung) including buying gold from them and being safely guided through the bush. In this picture Rowe depicts the multi-cultural nature of goldfields' society, including a Jarra/ Djadjawurrung man on the far right.

'Parker and Macord, Potato Salesmen and General Fruiterers, Bendigo', Rowe, George, (ca 1857), watercolour. National Library of Australia, an1406538.

Many miners wrote of how they profited from joining with Aboriginal people on the Victorian goldfields, as reflected in this illustration where Aboriginal people are depicted with a plentiful supply of fresh food, a rare commodity on the goldfields.

'A party of diggers joining a native encampment', Read & Co, (1853), tinted lithograph, London: Read & Co. Rex Nan Kivell Collection, NK11291/C, National Library of Australia, an8930053.

Alcohol abuse on the Victorian goldfields was considered to be at epidemic levels and had particularly severe consequences on Aboriginal communities already reeling from the effects of dispossession of their lands.

'Diggers at a sly grog shop warned of the approach of a commissioner', Mason, Walter G, (1857), wood engraving, Sydney: JR Clarke. Rex Nan Kivell Collection, NK2106/112, National Library of Australia, an8003921.

The policing of the sale of illicit alcohol to Aboriginal people was difficult to enforce in towns and cities but was virtually impossible on the bush diggings such as depicted in this picture.

'Sly grog shop at Hanging Rock diggings', Mason, Walter G, (1857), wood engraving, Sydney: JR Clarke. Rex Nan Kivell Collection, NK2106/111, National Library of Australia, an8003926.

The cooperative relationship between Aboriginal and non-Aboriginal miners who were keen to secure traditional Aboriginal foods is a frequent theme which goldfields' artist ST Gill recorded in his artwork.

'Kangaroo stalking', Gill, Samuel Thomas, (1856), lithograph, Melbourne: James J Blundell & Co. Rex Nan Kivell Collection, NK2096/4, National Library of Australia, an7150060.

Gold was very much a great equaliser. Small groups of miners, often just pairs as depicted in this picture shared much adversity in their quest for gold. Miners from very different backgrounds sometimes formed strong bonds of friendship and co-dependency.

'Gold digger and Australian native', Unknown artist, (ca 1855), watercolour. Rex Nan Kivell Collection, NK6878, National Library of Australia, an6617514.

Visitors to the goldfields were often struck by the multi-cultural nature of the population on the Victorian goldfields. The exotic attraction of each other was often portrayed by goldfield artists.

'Australia – news from home', Baxter, George, (ca 1853), lithograph, ink on paper. Collection of the Australian National Maritime Museum, Image 00008756.

5. Trade, commerce and the service sector

The social constructions of Aboriginal people as workers have often represented them as poor or indifferent. A suite of writers has documented the significance of Aboriginal pastoral workers in northern Australia. However, only a handful of scholars have examined the extent and significance of Victorian Aboriginal people as a labour force in the nineteenth century and these have concentrated primarily on Aboriginal peoples' entry into the frontier economy of the 1830-40s. Yet evidence of Aboriginal entrepreneurship and employment in other less 'visible' yet instrumental occupations abounds during the gold rush period. This is important as it marks what historian Henry Reynolds described as a 'powerful riposte to the generalist historiography' which has relegated Aboriginal work, trade and commerce during the gold rush period to a 'desultory footnote'.

Aboriginal employment in the service sector, common in the squatting period, continued into Victoria's gold era with Aboriginal people acting in roles such as 'postie'. In the Orbost district Jack McLeod, an Aboriginal man carried 'mail for the Orbost and Corringie Stations'. Two unidentified 'Blacks' were employed as postmen in the Yea district and also in the Snowy River district. Alfred Joyce, a pastoralist in the Central Highlands of Victoria noted the employment of Aboriginal people as postmen, an event corroborated by famed gold artist, ST Gill, whose artwork 'The Bush Postman' portrays the scene described in writing by Joyce.

Aboriginal expertise at bark cutting (for building huts and water races) was renowned amongst miners. This type of commercial activity was a specialist activity in both their traditional economy and already established involvement in the pastoral industry. Some idea of how proficient and profitable the bark cutting trade was for Aboriginal people can be gleaned from the correspondence by the Bishop of Sydney, returning from a visit to the diggings around the Ophir region (New South Wales) in June 1851. The Bishop reported on the relative affluence of Aboriginal bark cutters compared to the white miners:

> Native blacks straggled in from the hills with their gins and picaninnies and received good pay for fetching firewood as well as bark for hut roofs. "Black fellow rich now" they said as they smoked cigars which many diggers could not afford. Riders gave the blackfellows their mounts to herd "Three shillin and tix pences, mind it horse," was the regular price. Troops of near-naked aborigines from the far outback trudged to look in wonder, display their skill throwing mulga boomerangs, stamp out their rhythmic corroborees, and beg gratuities.

The struggle for miners to earn enough to sustain themselves was often keenly felt on the Victorian goldfields as well. Gold seeking often yielded inconsistent and poor returns as attested by James Nisbet. Nisbet, relating his experiences of encountering Aboriginal people (presumably Wathawurrung) at Ballarat, did not describe them as mining, but (as in the Ophir region) noted their presence on the goldfields and that they were earning some money from gold diggers: 'met a party of half a dozen at Ballarat' who were sometimes 'employed by the diggers in remote gullies to strip trees of their bark for a hut, for a day's labour at which a little bread or a English shilling is sufficient recompense'. Similarly, Ned Peters' party at Dunolly employed some (presumably) Djadjawurrung to get 'some dozen sheets of bark for us' and who expressed their surprise at how many sheets the miners required, stating 'What for you so much like 'um hay? Piccaninnie wheelbarrow no good long with big one bark!' John Briggs, a Tasmanian Aboriginal attracted to the gold diggings with his wife Louisa, like many gold diggers, turned his back on gold digging and became an employee at Eurrambeen Station as he found it more profitable doing bush work such as cutting bark at 6d a sheet. William Thomas, the Guardian of Aborigines, reported that Aboriginal people were 'industrious and profitably employed' in cutting bark. In 1861 he wrote at length about Aboriginal businessmen and women brokering contracts and submitting tenders within an expansive building industry that was heavily dependent on bark:

> The Yarra tribe know how to work by contract. A case occurred up the Plenty two years back: a barn was erected, one Wonga was asked what he would charge for roofing it with bark. He went round the building two or three times, consulted the three blacks with him, and finally said "cut bark where we find good trees, *only cut it*, you cart it away, and white man put bark on, pay us black fellows two pounds." The same black made a contract with a publican in Richmond for the cutting of bark for the first shanty public-house on Anderson Creek Diggings.

Domestics

As Ian Clark and Diane Barwick have shown, many Victorian Aboriginal people throughout the nineteenth century chose to remain in their own country, stating quite categorically their desire to remain in the localities of their birth or their adopted country. In the midst of the intense labour shortages experienced in towns and cities precipitated by the alluvial gold rushes, Aboriginal peoples' cultural preference to not leave their occupations and home estates ensured the desirability of their labour. Aboriginal people too, it was noted, enjoyed or exploited the law of demand and supply. One white woman in Melbourne during the gold rush period exclaimed: 'I cannot at any price get a man to chop my wood, and l think myself fortunate if l can prevail on the black gins (natives) to work for half an hour'.

The employment of Aboriginal people as domestics or servants was commonly but briefly noted, predominantly off the goldfields. In Geelong, one correspondent wailed about all his servants leaving for the goldfields except 'a native black [Wathawurrung] to cook, and a native boy to wait at table & c ... one flock of 5,000 sheep under charge of a native black'. Christina Cunninghame, at Wanregarwan Station, near present day Molesworth, grumbled about having to 'do all the kitchen work with the help of a Black woman the only useful one of her tribe who is fortunately here at present'. Charles Panton recalled that during his stay at Mangalore he obtained the services of several Aboriginal people multi-tasking as guides, shepherds and rouseabouts and had also employed the 'chief as butler and his wife as char'. The relationships forged between non-Indigenous women such as Annie Fraser in the back country of Victoria and Aboriginal women were often vividly recounted as warm and memorable. Pastoralists in the Charlton district recalled that 'the lubras had kind hearts and helped the station women with housework and the children'. Squatters at 'Tandarooke', near Camperdown in western Victoria considered it was God's providence that the 'children of the desert' (Aboriginal domestics) had chosen to stay on their own country (and work for the colonisers) by comparing themselves to God's prophet Elijah being 'fed by the ravens in the wilderness when abandoned by your own countrymen and women'.

Sue Wesson has identified two Gippsland Aboriginal sisters from the Metung region, Elizabeth Thorpe and Emma Booth, who went to the goldfields, probably at Delegate, where they worked as cooks and laundry workers. It is highly probable that this was a scenario often played out on the goldfields as it had been in the pastoral era. After the alluvial gold rush period there is also evidence that a number of Aboriginal women were employed as domestics. Historian Kathleen Gannan discerned from newspaper reports and journals that Aboriginal people were predominantly employed as 'odd job boys, housemaids, guides and stock hands' in the Swan Hill region. In 1873, John Green, in correspondence to the Board for the Protection of Aborigines reported that two unidentified Aboriginal women were working for Mr C Reid at Reidsdale 'One of them is nurse in Mr. Reid's family, the other does all Mrs. Reid's washing'.

Babysitting

A feature of Aboriginal culture was the intense care of their children. Their attentiveness towards babies was reportedly also directed towards non-

Indigenous children. Women such as Katherine Kirkland in the Trawalla district of central Victoria recorded the Wathawurrung's adoring attitudes towards her infant daughter in the early pastoral period. The loneliness and isolation of non-Indigenous women on the goldfields was at times crippling, only obviated by the presence of Aboriginal women. This is exemplified by an observation made by a miner on the Omeo goldfields who noted: 'Tom Shehan had a young wife and child, she was the only white woman among hundreds of diggers and aboriginals'. There is an unverified suggestion that Wathawurrung people took care of the Eureka rebels' children at Black Hill in 1854 whilst they were at the stockade, but in any case there was certainly a great deal of such activity on and off the goldfields. At the Dunolly goldfield, Charles and Sarah Belcher received visits from Djadjawurrung people who gave much attention to their baby. When Charles Belcher went to Castlemaine for supplies, these Djadjawurrung people took care of his wife and child; they also cut wood and brought water. At Ballan, in central Victoria, vivid memories of Wathawurrung women crowding around Mrs Denholm's baby were recalled:

> Mrs Denholm was attracted by chattering human voices outside, and on going out, much to her surprise and alarm, found her baby, which she had left in its cradle, in the arms of a lubra, with several of the lubra's dusky companions crowding around it. The white "pickaninny" was the object of the greatest wonder to them; they kept touching and gently pinching its skin, as if they doubted its reality, and though Mrs Denholm was not satisfied until she had the child in her own arms, she realized that the lubra's intentions were kindly, and that there was nothing further from their thoughts than any desire to do the infant harm. As long as they remained about the place the baby was the object of their solicitude. They took advantage of every opportunity to see it, and the wonder it at first inspired never seemed to decrease.

An Aboriginal couple, Jimmy and Sally, in the Mitta Mitta district were known to 'be very fond of white children and often nursed Lucy Greaves [born 1862] when she was a little girl'. The Martin family too was very fortunate in having enjoyed baby-sitting of their children from the Wergaia and the Wathawurrung/ Djabwurrung.

> Mr. Martin was in business distributing groceries and he was away from home [Nhill] when his daughter was born. The Aboriginal women befriended Mrs. Martin and were very helpful at the time of the birth [March 1866]. They were particularly taken with the white baby and were very attentive towards her. Soon after Elizabeth was born the Martin family took up land at Lexton between the Langi Kal Kal Road and the main road to Springs. There was an Aboriginal camp near

the Toll Gate and one day Mrs. Martin discovered that Elizabeth was missing. A search was begun and some time later Elizabeth was found safe and well in the Aboriginal camp sucking a possum bone.

Such intimate encounters between families, though probably infrequent, would inevitably involve cross-cultural dialogues very significant for those mining families. Other family reminiscences also recall the warm solace received by non-Indigenous women from Aboriginal people. James McCann recounted how 'one old [Wathawurrung] lubra used to nurse me when I was a little fellow'. Solitary references such as 'many times Mrs. Baxter was left with only the blacks and her young family' may belie an intricate dependency forged between mining families, especially women separated from their miner husbands, and Aboriginal people.

Farming

During the Port Phillip Aboriginal Protectorate period (1838-1850), a protectorate station had been established on Djadjawurrung land near Mt Franklin in central Victoria. In 1855, demand for farming land led to a subdivision and sale of much of the Aboriginal reserve. Only 640 acres close to Mt Franklin were reserved for use by four Djadjawurrung families who wished to 'cultivate and sow the land which is indeed their own'. ES Parker, former Assistant Protector of the Aborigines, reported to a Government Select Committee in 1858: 'They hold twenty-one acres of land co-jointly … on their own account … They have erected decent residences for themselves; have cultivated the soil; have taken several crops [since 1852]'. Many miners and travellers passing through to the nearby diggings acknowledged that the Djadjawurrung farmers had moved quickly to grasp the economic opportunities and capitalised on the nearby goldfields by 'cultivating and selling produce' to miners at the local diggings. Some observers thought it notable that these Aboriginal farmers were 'in no respect different from ordinary European peasants in the habits and associations of their lives'. Illness, the accidental deaths of several of the Djadjawurrung farmers, encroaching non-Indigenous farmers, theft of stock by non-Indigenous miners and lack of land tenure security afforded by the government all spelled an end to the Djadjawurrung initiative.

Possum skin rugs

Similarly, the Aboriginal skills of possum skin rug making were readily employed by miners. Whilst many 'diggers were very fond' of hunting possums and 'making beautiful rugs of them, by sewing their skins together', miners and others more commonly accepted that Aboriginal people were more adept at the trade and thus engaged in what Edward Tame described as a 'good item of commerce'. Samuel Clutterbuck, a miner at Mount Alexander, described in some detail his trade in 'opossum skins for a rug' stressing that 'The present of skins was accompanied by the usual application of 'Give it flour, "Bit bacci" powder, lead, shot &c. &c.'. AB Pierce also extolled Aboriginal peoples' manufacturing skills, noting 'they are very adept in curing skins perfectly' which are 'taken into the townships for sale'. A number of miners were tutored by Aboriginal people in the method of possum skin rug manufacture. A Batey acquired his knowledge from another miner who had been taught by Aboriginal people:

> I learnt the art of curing skins from Gardner at least simply drying them after a fashion he had learnt from the blacks. Small wooden pegs are cut and the skin is stretched with them on the back of a tree and left a couple of days in the sun after which they are ready for use most frequently they are sewn together for rugs 50 or 60 making a covering more durable and much warmer than a blanket.

Owen Davies, a miner in the Ovens Valley, noted that the 'natives make a great use of their skin' whilst William Howitt observed that their trade and commerce extended past inter-tribal lines: 'they fish and hunt, make baskets and opossum rugs, and sell their produce to the white men'. Artist and miner on the goldfields, George Rowe, wrote that the 'opossum fur is beautifully soft and makes a warm covering to sleep under and is what most diggers have as it is very light a good one costs four pounds'.

Aboriginal people moved quickly to grasp the economic opportunities presented to them. F Hughes, a Castlemaine pioneer, recalled that possum skin and kangaroo skin rugs were 'sold to settlers and lucky gold diggers at five pounds a-piece'. Miner James Arnot bought a possum rug in Melbourne made of 72 skins sewn together with sinews, also for five pounds. Aboriginal people from the Mitta Mitta and Little River districts, to the east of the Ovens goldfield, paid regular visits with possum rugs for sale. 'Neddy Wheeler', an Aboriginal man from the Yackandandah region in the 1850s was widely known to trade extensively in 'valuable' possum skins and lyre bird tails for the millinery industry. The Indigenous manufacture of possum skin rugs, baskets and mats enabled many people at missions and reserves such as Coranderrk, Lake Condah (Western district of Victoria) in December 1870 to gain an eagerly sought-after economic independence. Reports from a number of Aboriginal Station managers across Victoria describe the lucrative trade being conducted. John Green, the manager of Coranderrk, wrote: 'In the course of one week or so they will all be living in huts instead of "willams" [traditional bark housing]; they have also during that time [four months] made as many rugs, which has enabled them to buy boots, hats, coats etc., and some of them has [sic] even bought horses'.

Green and other mission managers reiterated the ready sale 'at high prices of baskets ... rugs with the skins of the opossum, kangaroo and wallaby, for each of which they get from 1 pound to 1 pound 15s'. Kulin people at Coranderrk increased their production of 'rugs and baskets' to such an extent that by 1868 the sales of their Indigenous manufactured goods represented over 20 per cent (100 pounds) of the annual value of production of the station.

It is hard to determine exactly why Aboriginal entrepreneurs participated in the fur, feather and skins trade with the colonists, but presumably the acquisition of guns, exotic foods, tobacco, alcohol and other western goods were a strong incentive. In John Zwar's boyhood reminiscences of the 'Puckapunyal tribe' in the 1860s, he recalls their business-like adroitness when selling their artisan products to non-Indigenous townsfolk:

> The Puckapunyal tribe, about twenty in number, made baskets out of rushes ... The blacks made very fine baskets out of the rushes and sold them to the people. I do not remember the price charged but I know they refused brown money; it had to be white [silver or gold].

It is not difficult to determine why non-Indigenous miners sought to purchase the Aboriginal-made rugs. Miners considered that one rug imparted as much warmth as a dozen blankets. The importance of possum skin rugs for the miners is exemplified when in 1865 a miner in the Carngham district, Henry Davies, sought to get the local Guardian of Aborigines to 'get an opossum rug made for him, to take home to the old country, to show what the pioneers of the goldfields frequently used to sleep in. An Aboriginal couple was engaged to make a rug that they completed in four days, and were paid 30 shillings'.

The sale of possum skin rugs, baskets and artefacts to non-Indigenous miners, as with the commodification of corroborees for non-Indigenous audiences, had commenced from the first days of colonisation in Victoria, 15 years prior to the gold rush. During the squatting period, inter-cultural economic activity between Victorian Aboriginal people and non-Indigenous people was enacted for social, political and monetary purposes. But increasingly during the gold rush period the emphasis in inter-cultural trade focused on financial gain. Aboriginal peoples' 'commercial instinct' was noted by miners to be highly developed, meaning they knew the monetary value of their manufactured products and asked a good price for it, yet embedded within such commercial transactions were probably the systemic values of reciprocity and kinship. Samuel Clutterbuck noted with some displeasure how money had clearly displaced barter in Aboriginal dealings with non-Indigenous people:

> The blacks took their departure, Simon promising on his next passing, to bring me a new opossum rug and one each of their different implements

of war and hunting. I asked him if I should give him a fine shirt in return. He replied "Borag [Borak: No] shirt, give it plenty white money". I may here state, that the "amor munni" [love of money] is as strong with the aborigines as their paler faced brethren.

R Brough Smyth, a noted nineteenth century ethnologist, also considered that Victorian Aboriginal people 'barter with their neighbours; and it would seem that as regards the articles in which they deal, barter is as satisfactory to them as sale would be. They are astute in dealing with the whites, and it may be supposed they exercise reasonable forethought and care when bargaining with their neighbours'. Walter Bridges, a miner at Buninyong near Ballarat in 1855, described how a local clan of Wathawurrung people carrying possum skin rugs approached his wife and made a request, framed within the ties of reciprocity of neighbours, for some steel needles and thread:

> So up they come yabbering good day Missie you my countary [sic] woman now. My mother had to be the spokesman the Blacks said You gottum [sic] needle Missie you gottum thread … Then the Luberes [women] come jabbering along behind carr[y]ing the swag in nets some with pups that could not walk, others possum skin rugs the Blackfellows make.

It is probable that the demand for non-Indigenous sewing implements stemmed from the high volume of possum skin rugs being sold on the goldfields.

The volume of trade in possum skins increased exponentially. Edward Tame, a traveller on the goldfields, noted that the skins of possums 'form good articles of commerce' for the 'Aborigines'. Newspaper reports from home and abroad also reveal a strong interest in Indigenous manufactured goods, particularly in possum skin rugs. In 1861 the *Ballarat Star* carried a satirical article attributed to 'A Blackfellow' which beseeched the colonial government to provide market protection for the Indigenous trade in possum skin rugs:

> You write guv'nor and ask him why protection on the wallaby track looking for grubs 'mong whitefellow? You say whitefellow no make um blankets this colony, blackfellow make 'possum rug, which whitefellow ought to buy 'stead of blanket; possum rug all along same as whitefellow's blankets;- why not give blackfellow monopoly of making and selling 'em and protect real native industry.

Eugene von Guérard, artist on the Victorian goldfields, documented such a trading transaction in 1854. His oil painting, 'Aborigines on the road to diggings' or 'The Barter', now in the Geelong Fine Art Gallery, depicts Wathawurrung people offering possum rugs for sale to white miners on their way to the goldfields. Of particular interest is the centrality of the Wathawurrung men and women. Unlike many depictions of Aboriginal people during the nineteenth century, peripheral players cast off to the background or figures relegated to the sidelines, Von Guérard has focused the activity around confident Aboriginal salespeople who are clearly directing the business deal. The white 'consumer' desiring to purchase the possum rugs is painted in a subservient pose, kneeling down, whilst the Aboriginal 'manufacturer' assumes an upright, dominant demeanour.

Corroborees

Ceremonies performed by Victorian Aboriginal people, described under the banner of 'corroborees', were performed for non-Indigenous people from the very outset of non-Indigenous contact. The Melbourne-based magazine *Table Talk* commented on 21 January 1887: 'Ever since British rule was established in Australia, an aboriginal "corroboree" has always been considered an amusing, if not a particularly edifying spectacle for distinguished visitors'. The popularity of the corroboree as a piece of theatrical entertainment was immense in both the pastoral and gold period. Though not often articulated by historians, the nineteenth century corroboree performed for non-Indigenous colonial audiences was Australia's pre-eminent prototypical Indigenous cultural tourism product. The development of the corroboree event by Aboriginal (and non-Indigenous) entrepreneurs gave witness to some very innovative and successful transformations during the gold period. There is a considerable corpus of evidence demonstrating the gold period was a catalyst for corroboree performances. JD Mereweather, for example, was informed that the corroboree he witnessed in central New South Wales 'had come from the coast of South Australia'. Similarly, Lawrence Struilby witnessed corroborees that had been passed on with greater rapidity since the gold period had opened up communications in the interior. One respondent to the 1858-59 Select Committee on Aborigines related how he had once 'traced a song which I knew to have been composed at a particular time near Port Stephens, and found that in the course of about three years it had been brought down through Bathurst, Yass, the Murrumbidgee, and the Murray, to Melbourne'.

Many residents in the Ballan and Bacchus Marsh district of central Victoria recalled corroborees being held by the Wathawurrung close to the townships and other locations during the gold mining period. There were also writers who penned their perceived notions of corroborees into their reminiscences to afford some savage exotica to their tale of the goldfields.

In their description of corroborees, miners such as Edwin Middleton emphasised the vigour, dramatics and nudity of the performance. Both impressed and repulsed, Middleton wrote:

> At night they had a big Corroboree on a grand scale: 4 women sitting by
> a fire beating time with sticks and 8 or 9 men dancing the Corroboree.
> With the exception of a small apron made of grass, they were entirely
> naked. As they danced they kept up a humming noise, every now and
> again breaking out with their cry of Coohooee. The men were daubed
> all over with coloured clays and looked most hideous. To see them every
> time their Gins put fresh leaves on the fire, you might fancy you were on
> the borders of some of the [   ] regions.

Charles Panton, a Commissioner on the goldfields, appreciatively described
several different corroborees performed by hundreds of Aboriginal people
'night after night' at Mangalore Station in Central Victoria. Samuel Clutterbuck
told of being 'summoned' to see a corroboree and during a break in the
festivities, being asked for 'bacca' which signified an exchange had occurred.
George Admans recollected that Wathawurrung people, after receiving supplies
on behalf of the government from the lighthouse keeper at Queenscliff, Victoria,
were 'accustomed to remain for a few days and before leaving to entertain us
by giving a corroboree' in the 1860s. Other observers such as William Simkin
recalled similar events also at Queenscliff in the 1860s: 'These blacks used to
entertain the visitors in the day by throwing their spears and boomerangs,
and in the evenings, by what they called a Corroboree'. Simkin also alluded
that there was an element of conscious ritual exchange occurring between the
respected non-Indigenous 'elders' and the Wathawurrung head clans man:
'Some of the gentlemen visitors would give the king of the tribe a part cast off
dress suit of clothes namely a swallow tail coat and belltopper hat, and after the
performers had sung and danced, the king would go around collecting with his
hat ... they were liberally supported'.

Michael Parsons, writing on the tourist corroboree in nineteenth century South
Australia, has argued that corroborees staged for a non-Indigenous audience
emerged as a 'cultural product jointly negotiated between two cultures'.
Furthermore, he posits that during the process of negotiation, four major
framings of corroboree can be identified. First was the 'peace corroboree',
marking a new state of cooperative relations between Aboriginal people and
the Crown, or representations of the Crown such as 'gentlemen squatters'. Thus
RG Jameson (1852) related how a recently arrived family to the Colony had
received an invitation from 'King Jack, the chief and his wife' who 'conveyed
to us the information that a large native gathering and corrobor, or dance was
to take place with "plenty of noise" that evening; but that no harm need be
apprehended by his white friends'. Second, the 'command performance'
corroboree, was orchestrated by the new occupiers as a joint act of homage
to the Crown or other significant notables. This was itself a re-framing of the
corroboree as a traditional act of welcome, but also functioned for settlers as a

handy piece of ready-made, uniquely local, pageantry that could be included on the program for notable official visitors. Thus, in 1867, Buninyong Council wrote to Andrew Porteous the local Guardian of Aborigines in the Ballarat district informing him 'the Council has determined on getting up a grand corroboree of the Natives on the occasion [his Royal Highness the Duke of Edinburgh's visit to Buninyong]. I have therefore to beg that you will be pleased to secure as many of the aboriginals as possible for that purpose; every care will be taken of the Blacks whilst in the locality'. Third, the 'gala' corroboree, marked significant social occasions. The intermingling from one framing to another allowed any significant colonial or Aboriginal occasion – anything from the opening of a railway line to the separation of Victoria from New South Wales – to be accorded a vice-regal relish.

Parsons described the fourth type of corroboree as 'touristic'. This type of framing of the corroboree is particularly evident in goldfields records. In the Minutes of the Ballarat Mechanic's Institute in February 1879 are references to the hire of a lecture hall for an 'Aboriginal Concert', presumably an instance of non-Indigenous goldfields promoters trading on the sense of exotica which Aboriginal performances provided. A corroboree orchestrated by Wathawurrung clans (probably Carninje balug and Wongerrer balug) amply illustrates that as access to their land and its raw materials – their economic capital – was progressively denied them by pastoralism and then gold mining, Aboriginal people seized upon the opportunity to market their cultural knowledge and skills – their symbolic capital – and convert it, not just into hand-outs of food and tobacco, but hard currency.

> A CORROBOREE – During the past few days the town of Smythesdale has been infested by a numerous gang of aborigines-men, women, and children. On Tuesday and Wednesday they went about the town in quest of sixpences, tobacco &c., and announcing a grand "corroboree" to come off on Wednesday night, as it accordingly did, in the presence of a hundred spectators or more. The savages were in their war paint, and looked sufficiently frightful as they danced and shrieked round their fire. The scene of the orgie was in the wood over the creek, near the Carngham road; and the dissonant noises, vocal and instrumental, which formed part of the entertainment, were distinctly heard at the firesides in the township. The thing was kept up till an advanced hour in the morning.

This event and others like it, independently organised and without joint partnership of any kind, were pre-planned. It is plausible that in one sense the gold era ushered in more opportunities to perform corroborees, as more spectators inevitably ensured more food and rewards for the performers, and the application by non-Indigenous people for more performances may have led to some small degree of reliability of income. Tommy Anthony, for example, would

> attend football matches [in the Wangaratta region during the latter half of the nineteenth century] and give demonstrations of boomerang throwing. Afterwards he would go around with his hat. It is said that for the first few throws he would attempt to catch the boomerang when it returned, but purposely missed. A few more pennies would be thrown to the ground in a wager and when he considered that there were enough there, or someone might catch onto his game, he would make the boomerang soar and deftly catch it on its return.

Non-indigenous ceremonial occasions such as Christmas Day may have been taken advantage of, judging by a diary entry by Mrs James Madden, of corroborees in the Ballarat district in 1853 which were well attended by Aboriginal (possibly Wathawurrung) and non-Indigenous miners: 'Ballarat seemed to be on the wane and we set out for Mt Cole for timber for the homestead at St Enochs … Here we arrived on Christmas day, 1853, and were entertained by about two hundred and fifty blacks at a grand corroboree at night'. Reports such as appeared in the *Corowa Free Press* in 1876 indicate that Aboriginal people were financially supporting themselves through performing corroborees, and were adept at petitioning for pre-booking fees, actively promoting and marketing their cultural heritage product:

> The Aborigines had promised for some weeks beforehand that there would be a grand affair on a Saturday evening which was well attended by the townsfolk … some of them going around town asking for a mug of beer or some food, or some cast-off clothing before the corroborees … At the conclusion the hat was taken around and the ringmaster stated: 'That he didn't mind if they gave one shilling or six pence Koonong'. Tommy said they would have an ever greater corroboree when the Wangaratta blacks came over.

In addition to Parsons' four major framings of corroborees or dance events staged by Aboriginal people for a non-Indigenous audience could be added the 'solemn and sacred' corroboree encompassing purely Aboriginal matters that was strictly by invitation only. William Thomas, Assistant Protector of Aborigines in the Melbourne and Western Port districts was advised by a Boonwurrung elder in 1839 that some corroborees were off limits to white people, being explained as being too sacred and likened to 'white fellows Sunday'. There is some evidence that during the gold period 'invitation only corroborrees' were still practiced. Henry Burchett on the Lower Loddon diggings wrote in 1854 of being privileged to attend a 'native funeral ceremony to which few white men are admitted …

after asking permission'. A satirical *Argus* report of an impending corroboree by Wathawurrung clans in the Geelong district reprinted below, and in the same year a report of a very large gathering of several language groups at Braidwood, indicates that *Tanderrum* or 'welcoming to country' ceremonies were still being performed, but increasingly under the shadow of the coloniser's gaze. Events such as these were probably viewed by non-Indigenous people who may have been barred from attending such events prior to the gold rush period.

> FASHIONABLE ARRIVALS – His Royal Highness Ko Ko Warrion, King of Colac, arrived at the Barwon Bridge a little after ten o'clock yesterday morning, from whence His Royal Highness was escorted into town by a mottle assembly of young blades, who did everything in their power to make the King's entrance into the ancient city of Geelong a perfect jubilee. His Majesty's suite consisted of nine peers and a countess. We were given to understand by one black fellow that Geelong was honored by the royal visitors for the purpose of dancing "great corroboree" this day, in honour of our most gracious Majesty's birthday.

Some goldfields commentators were aware that corroborees were not purely performed for amusement but also were for religious and martial purposes. Ceremonies were performed for goldfield audiences which had profound spiritual significance and educational importance for Aboriginal people. In May 1860, for example, 'an extraordinary exhibition' was given by the Djadjawurrung at Lamplough mining community. The educated members of the non-Indigenous audience may have recognised that the exhibition was a dramatic re-enactment of the Djadjawurrung's belief in spiritual reincarnation. One scene involved two men decorated with white chalk killing a third man, painted red, and burying him under one of the stage's trapdoors, whereupon he reappeared from another trapdoor smothered with white chalk, having 'jumped up whitefellow'.

The religious or 'peace' corroboree was also recorded as being performed on the goldfields, probably witnessed by uninvited non-Indigenous audiences. George Rowe, writing a letter from the Castlemaine diggings recorded a conversation with a Djadjawurrung woman who had remonstrated with a neighbouring clan and intimated that a peace making 'religious' corroboree was to be held at the McIvor diggings.

> The group was the Bendigo tribe two men and a woman came up while I was at there and a young woman addressed them in a rage and threatened them with a stick to drive them away one of them a very fine handsome fellow with a slightly aquiline nose standing erect above 6 feet stood calmly surveying them all the time with his long spear erect – she got tired at last when she told me she spoke a little English "that he Murry river blackfellow he kill Bendigo blackfellow he no here Bendigo he go

away" – after all she gave him a loaf of bread and another party gave him half a damper which they had been eating from and then they walked they together with the Bendigo tribe were on their way to MacIvor to a corroboray on the next evening being new moon it is some sort of religious ceremony when they dance all night and a very large number arrive perhaps 500 or 1000.

It is quite likely that the gold era ushered in, or at the very least, enabled a greater frequency of, the staging of corroborees just for the fun of it. Following the discovery of a 'four and a half ounce nugget' by Aboriginal people, a reporter for the *Maryborough Advertiser* noted: 'The proceeds, amounting to 18 pounds, were soon disposed of, and a grand corroboree has been held ever since, and doubtless will continue till all the money is gone'.

Goldfields newspapers tended to memorialise the corroboree, which often was the subject of some commentary such as an article which appeared in the *Inglewood Advertiser* on 2 May 1865:

Natives. Inglewood is now honored with the presence of a body of natives, male and female, who have come down from their native river, the Murray, to see their white brethren inhabiting this town. It is some time since they paid us a visit and their appearance in the street, with their long spears, opossum cloaks, etc form quite a novel feature. There are about twelve of them, accompanied by some lubras and their King with a few picaninnies. Disdaining the use of the white man's road, they struck direct from the Murray through the scrub to Thompson's Gully, where they camped on Saturday night, right upon the spot where, six years ago a lubra was buried. Yesterday they were busy levying contributions, and showed a great predilection for white money. Last night they held a grand corroboree, which was quite a success, and attracted many visitors.

Without the aid of bill posters or newspapers to promote their event, it seems that on many occasions Aboriginal performers put on an 'appearance' with all their accoutrements in the streets to elicit interest in their exotic difference and announce their intent to perform, as described in this article from 1866.

Several corroborees have been held during the past week close to Echuca by the blacks, who, to the number of about 200, have assembled there from all quarters within a radius of about a hundred miles.

The Riverina Herald reports 'that there are a large number of fine, strapping and even handsome, young men among them-one fellow in particular, is notable for his height, standing on his bare soles over six feet. A young lubra, rejoining in the name of Polly, struck everyone who saw her yesterday as she perambulated High Street with her rather diminutive spouse, from her great height and erect carriage. We understand that she measures over 5ft 8in. There is a good sprinkling of old and very grey-bearded men ... The occasion is considered one of high holiday and festival. No doubt the time is spent idly enough, but as to the feasting, we fear it is not to be compared to what it was in the olden time.'

A partial glimpse into the Aboriginal outlook on the merits of touristic corroborees being performed in the gold rush period (partial because the Aboriginal discourse is viewed through non-Indigenous reporters' lenses) can be attained from newspaper reports. Glimpses of disappointment can be found in the *Ballarat Times,* where a correspondent, after witnessing a well attended corroboree at Lake Wendouree met with a small party of Wathawurrung people who 'seemed grieved at the revelry and debauch which on hands surrounded them, and was evidently taking no part in the noisy performance'. The disdain which this group held for such an event is juxtaposed with the renewed sense of independence and pride observed of a neighbouring Djabwurrung clan at Back Creek who performed for diggers also, although reported in a mocking manner:

> In one communication lately inserted in the *Star,* l stated that a party of Aboriginals had made their appearance histrionically, on the stage at the Royal, and that the audience were highly pleased at witnessing the intelligence of the sable tribe. The warm reception they got has completely transmogrified them from slow motioned bush wanderers to aristocrats, possessing strong self esteem. They were engaged to give a grand performance at the Back Creek Theatre Royal last week. The contractors no doubt expected them to walk the distance of ten miles, but he found to his astonishment that they had grown so aristocratic since their appearance at the Royal that they refused to stir one inch unless their passage was paid by Cobb's coach or a "special" grand conveyance procured for them.

Indigenous entrepreneurship

Aboriginal peoples' business savvy, which developed rapidly in the economic climate of the goldfields, can be seen by examining a number of newspaper articles which charted their economic and social empowerment in colonial society.

> On this particular instance the performer, once he had changed his dress, would go round to the visitors and make a strong appeal to each and sundry to give "black fellow a shilling" … The whole scene was one which, once witnessed, is not easily to be forgotten.

Not surprisingly Aboriginal people also chose to perform for non-Indigenous audiences in a range of goldfield venues including hotels, mechanics institutes and outdoor public meeting places.

> A most novel scene l witnessed at the Royal theatre on Thursday evening. The Ararat tribe of Aborigines [Djabwurrung] has been here for some days, and most pleasing it is to see them so far advanced in civilization. The women have their hair neatly combed and oiled and the men are dressed as Europeans. The King wears a white bell topper, of which he seems as proud as if he wore the Crown of England. An offer was made to them to appear and dance at the Royal, which offer was accepted with avidity. Upon the curtain being raised the dance commenced; and the strict time kept, together with their various steps, completely astonished the audience. After the first piece was over, one of them appeared at the footlights and announced a programme of what would be exhibited before us. In his intelligence and manners he was a pattern to hundreds l have seen of Europeans attempting to address an assemblage. One of them has gone up the country for fifty more, and a grand evening's entertainment is to be given by them at the Royal on Saturday evening next.

There is evidence too of both traditional rivalries between distant language groups being relaxed and of Aboriginal people metamorphising the traditional coming together component of corroborees to include a commercial arm to the proceedings, as evidenced by a report of a gathering at Ballarat in March 1861:

> During the last few days a number of aborigines, probably about two hundred, have arrived on Ballarat from Port Fairy, Mount Elephant, Mount Cole, the Hopkins, Warrnambool and the Wimmera, for the purpose as they state of seeing the towns and each other … During the whole of Monday they infested the principal parts of the town and levied contributions in money or otherwise on the white man. Towards evening they made preparations for a corroboree in the Copenhagen grounds … and were a considerable time in getting the music to a proper pitch … Steam however was got up at last, and away they went to the intense delight of some 500 persons, who were present to witness the performance … While the dancing was going on King Wattie procured a tin can, and fulfilled the not very dignified position of tax-gatherer

in-chief, but up to nine o'clock he did not appear to have been very successful in inducing the invader to acknowledge his right to impose taxes when he liked.

An advertisement (20 February 1865) and subsequent news report of a 'gala' event printed in the *Ballarat Star*, reveals that 'touristic' type corroborees were carefully planned and intended to utilise Aboriginal peoples' heritage as a vehicle for economic self sufficiency.

COPENHAGEN GROUNDS

Grand Corroboree by Fifty natives

THIS EVENING, MONDAY, 20th INST.,

Also, Extra Exhibition of FIREWORKS and Balloon Ascent. For the Benefit of Professor Prescott. Grandest Gala Night of the season

The aboriginal corroboree and display of fireworks at the Copenhagen grounds on Monday evening drew together a large number of persons, and the novel entertainment proved a decided success. Aboriginal habits in their most primitive style were displayed by about thirty-five natives, from various tribes around Ballarat, including about a dozen lubras, who were nearly naked and daubed over with paints of every hue in the most hideous fashion, though no doubt after approved aboriginal style. Without offering any comment upon the propriety or otherwise of the corroboree, it may be stated that it afforded amusement to the number of persons, between five and six hundred, who assembled to witness it. A plentiful supply of coloured fires added to the savage appearance of the scene, and after it was concluded some beautiful fireworks were displayed. Professor Prescott, the lessee of the grounds, purposes on a future evening to allow the natives the use of the grounds for another corroboree, they receiving the proceeds.

These Indigenous theatrical cultural products (corroborees, boomerang throwing, demonstrations of weapons, and so forth), jointly negotiated between two cultures, were increasingly instigated by Aboriginal people for monetary purposes. Adopted and adapted within the cultural parameters of traditional Aboriginal reciprocity and kinship, they reveal a level of cross-cultural convergence along economic lines beyond anything previously documented (or imagined) by historians.

6. Co-habitation

Historical records relating to the alluvial gold mining period, predominantly from the 1850s, implicitly convey a degree of co-habitation between Aboriginal and non-Indigenous people. In his reminiscences, DJ Ross provides a representative example of influential early bush life on a squatting station (the first wave of dispossessors): 'Born on a sheep station my earliest recollections are in connection with blackfellows, bushrangers and shearers'. All across Victoria's gold mining districts Aboriginal men, women and children came into contact with non-Indigenous men and a few women and children. The degree of close contact between the races was variable, and differed in nature from friendliness to outright hostility. Relationships were complex, varying not only geographically but rapidly changing over time as one rush was superseded by another rush and extremely large heterogeneous and transient populations waxed and waned across the region.

Aboriginal voices

A belief amongst Victorian Aboriginal people in claiming certain non-Indigenous people as resuscitated clanspeople continued during the gold rush period. Samuel Clutterbuck recounted being instructed on this subject: 'I told him of poor Wight's death. Aha! Said he [Murray, an Aboriginal] "Mr Wight, quamby [stay or sleep] alonga this, (pointing to ground) come up black fellow, bye and bye." This is their tradition of the final state of white men and vice versa of their own people'. Clutterbuck further related another instance of a very dark skinned squatter 'who was suddenly embraced on one occasion by a black, who in great glee exclaimed "Brother belonging to mine, I believe quamby a long time ago on Murray"'.

Often the newly appointed 'brother' or 'child' did not understand the relationship and commitments that were being invoked but treated it as a friendly nicety to be humoured. Lawrence Struilby, a goldfields traveller, related a story of an unidentified Aboriginal clan who believed a whiteman who arrived in the vicinity with the same peculiar bent arm was one of their deceased – 'a blackfellow jumped up white fellow'. The clan subsequently 'would do anything for him and ... carried tons of split timber and bark to build his huts, &c'. This understated response from the new wave of colonisers in the gold period is particularly evident in Hubert De Castella's conversation with an unidentified Aboriginal man: '"You are my brother long time dead", one of the old men used to say to me with a sort of respectful friendliness'. In seeking to explain why Victorian Aboriginal people continued to hold this belief after many decades of contact with non-Indigenous people, it seems apparent that a type of synergy between Aboriginal spiritual beliefs and elements of adopted Christian beliefs might have been occurring amongst some Aboriginal people which in fact strengthened the traditional belief in reincarnation. Robert Hamilton, a Presbyterian evangelist who had had a long association with Coranderrk (Aboriginal Mission north-east of Melbourne) wrote of Wildgung (Old Jemmy Webster), a Taungurong Elder who had to all appearances accepted many Christian teachings because he believed the superintendent of Coranderrk, John Green, was his deceased brother reincarnated.

Most non-Indigenous people did not discern the full implications of the resuscitated kin relationship but assumed that Aboriginal people were 'honouring' them. Typical of this response is an entry in the diary of a son of a clergyman in Gippsland in the 1850s. He recalled how an Aboriginal youth 'elected to stay with us' and 'took to himself the name of Billy Login, in honour of my father'. The kinship association was continuously affirmed by 'Billy Login' in later years as he 'always recognized our family whenever met as "sister belonging to me," or "brother belonging to me"'. A small number of white miners recognised the level of importance that Aboriginal people placed on forming relationships with individuals, maintaining links with their land and subsequently reaping the benefits. One goldfields writer noted:

> As their services are given more from goodwill than from hope of reward, it is only from attachment to persons with whom they are well acquainted that they are ever prevailed upon to lend themselves as parties in an exploring expedition ... Old Bill Cowpers never seemed to move from the place where he first commenced. Perhaps it was very inconvenient for him to shift, as he had an aboriginal woman living with him, which might be a potent reason for his always remaining at one place ... washing with the assistance of the aboriginal woman ... when he came upon a rich spot he had got six hundred pounds' worth, [which] would keep him and his gin a long time.

When Aboriginal voices are articulated in goldfields historical documents (usually through the filter of the colonisers' writings) they are seldom ones of diffidence, especially in relation to their individual and collective rights. In both the pictorial and written records there are exemplars of Indigenous resistance to unfair governmental controls on and off the goldfields. The written records reveal Aboriginal people linking their grievances with the wider demand for civil rights and entitlements:

> A group of Aboriginal diggers at Forest creek in 1852 when asked to show their licenses replied to the mounted police that 'the gold and land were theirs by right so why should they pay money to the Queen?

Aboriginal people also attempted to invoke the traditional democratic practices of sharing their country and resources with visitors. Joe Banks, a Kurnai man from the Brodribb district, expressed his indignation to Constable Hall, a non-Indigenous Police Officer, after the Constable had failed to come to the aid of Banks and perform his duty:

> At one time Joe and his gin were camped out at Bete Bolong. A big flood came down. The old gin died. Joe sent word in to Orbost by John Johnston to tell Const. Hall to come out. Because of the flood, he did not come out for three days. Joe was very angry. "I will report you Mr. Hall, you should have been here three days ago." Mr. Hall said the flood had held him up and he would not be able to take the body to Orbost. The policeman suggested that she should be buried there. Joe agreed after a lot of protesting but he made Const. Hall dig the grave. "No that's your job."

It is almost certain that many gold miners' accounts of Aboriginal peoples' insistent *begging* were not desperate ploys to extract food and goods from white colonisers; rather, viewed as part of a relationship of exchange, were probably an expectation of being recompensed for use of land; or rent. The miner James Madden and his wife remarked on the ire of Aboriginal people, presumably enraged by the invasion of their homelands by hordes of immigrant miners in 1852. En route to the goldfields of Ballarat they met up with 'a big fellow who proudly assumed his kingship by stepping out to threaten us if we did not leave his terrain'. Bulmer, a missionary in the Gippsland region during the latter half of the nineteenth century, reported that he had 'great difficulty in persuading them [Aboriginal people] to work for what they get. They have an idea that I have no right to ask them to do anything in return for the food I issue; and many of them have gone away from the station'. In a like manner Wathawurrung workers berated Mr Young, the Honorary Correspondent for the Bacchus Marsh district, about the Aboriginal Protection Board's lack of deference to what was the right behaviour. He reported 'They have assisted this year in digging potatoes, and have been paid in cash and butcher's meat; the latter, they often remind me, should be furnished by the Board'. On the goldfields too Aboriginal disquiet about the perceived lack of sensitivities was voiced. A miner and his companions at the Bendigo diggings in 1854, who had camped near a newly made Djadjawurrung grave, were curtly told to make other arrangements as they had impinged on sensitive mortuary arrangements.

At times Aboriginal people tried to educate their non-Indigenous neighbours of their obligations and responsibilities. A Welsh farm worker noted in his diary that he had been given a scathing lecture by a Djadjawurrung man about the sub-standard etiquette and morals of white people:

> A dark native, that is an Aborigine, paid me a visit. He was looking for bees. He mentioned that when a native discovers a hive, he invites the neighbours to partake of the honey, but when a white Christian discovers it, he keeps the produce for himself.

Richmond Henty recounted a Justice of the Peace being sharply rebuked by an unidentified Aboriginal who when sentenced to be put in the lock-up for theft remonstrated: 'what for you say I steal? What for you steal my country? *You* big one thief! What for you quamby [camp] along o' here? Geego along o' your country, and let blackfellow alone'.

'I am the owner of the land about here'

Aboriginal people across Victorian goldfields continued to declare their title and insist upon formal acknowledgement of what was rightfully theirs. At an inquest held in 1870 a Djabwurrung Elder affirmed: 'I am the owner of the land about here. I was born near here. I am Chief of my Tribe – Chief of Buangor'. Others such as King Billy, 'the last of the Loddon [probably Djadjawurrung] tribe proposed in 1872 to erect a toll gate on the new bridge over the Loddon by "the right which his progenitors enjoyed in the ages of antiquity"'. Similarly, the churning of the Murray River which interfered with their supply of fish prompted local Aboriginal people to ask Philip Chauncy, a surveyor in the Goulburn district in the 1860s, to appeal to the government for a bounty to be placed on river vessels payable to the Aborigines. Chauncy wrote:

> A native of the Moira tribe informed me of the intention of himself and five other aborigines to proceed as a deputation to His Excellency the Governor, to request him to impose a tax of 10 pounds on each steamer passing up and down the Murray, to be expended in supplying food to the natives in lieu of the fish which had been driven away.

Many accounts attest to Victorian Aboriginal people assuming their traditional cultural entitlements. Charles Fead, a miner in the Buchan area, remembered a 'brawny aboriginal walking into the hut and helping himself to a drink of water'. JF Hughes at the junction of the Devil's river with the Goulburn described a similar scenario whereby two female aboriginals 'entered the "mia mia" [hut], minutely examining the contents, and after satisfying their curiosity and their wants, departed as mysteriously as they came'. The *Geelong Advertiser* reported that the Wathawurrung elder 'King Jerry proclaimed "his intention of demanding restitution [from Geelong City Councillors] of all provinces of which he has been illegally deprived, after having held them by indefeasible title from time immemorial, together with all improvements thereon, and revenues accruing from all sources"'. Some Aboriginal people, such as Equinehup, a Djadjawurrung man, formally petitioned colonial authorities (Railway Commissioners) expressing his claim to original land title:

> Gentlemen and brothers too, I am the last of the Aborigine tribe in these parts. I do Humbly wish you to compare two lots of title deeds. I received mine from the author of nature While the land occupied by all the railways Is titled by the white mans lawyers.

Andrew Porteous, the Honorary Correspondent at Carngham (near Ballarat), advised the Board of the local Wathawurrung clans' keen intentions to obtain some land for themselves on their traditional lands, stating that 'A number of the tribe have requested me to apply to the Government to reserve a block of land near Chepstowe for their use, where they might make a paddock, and grow wheat and potatoes, and erect permanent residences'.

Very occasionally it is possible to hear some semblance of Aboriginal voices or opinions about the gold rush and their appraisal of what was occurring on their clan estates. Some Aboriginal people were reported as expressing their disdain for the 'new' influx of outsiders on their land. John Moore, a miner at Bendigo, relayed how Djadjawurrung people expressed their distaste, not necessarily at the practice of mining itself as they were familiar with resource extraction, but at the fevered frenzy and psychologically disturbed character of the 'whitefellow all gone mad digging holes and washing stones'. A Djadjawurrung farmer at Franklinford in central Victoria frankly confessed in 1856 that 'for a time, at first, he did not like either Europeans or European customs'. Unfortunately, he did not divulge details of why he disliked Europeans, but it would not be difficult to hazard a guess considering what devastations had been visited upon the Djadjawurrung and their country. The disdain which Aboriginal people held for certain non-Indigenous classes of people was also occasionally aired. Anne Meredith at Mt Elephant in the Western district of Victoria (circa 1850s) made mention of how 'an aboriginal native who had for some time installed himself among the hangers on at our station, looking with an air of lofty contempt upon some of the new-comers, inquired of their master what he would possibly want with those [non-Indigenous] "wildfellows"'. Being spoken down to was particularly resented as JC Hamilton recalled.

> I remember meeting a young black woman in the early fifties, who was with the tribe in our district, but who had been for a short time at a mission station in South Australia. I spoke to her in pidgin English, and her answer was, "You need not speak to me like that; I understand English as well as you."

In answer to questions posed by the Victorian Government's 1858 Select Committee into the Present Condition of the Aborigines, Mr Hull, a District Magistrate, related a conversation he had had with Derrimut, a Boonwurrung elder, in what is now present day Melbourne which dramatically depicts the disempowerment and disenfranchisement that was keenly felt by Aboriginal people especially during the gold rush period:

> The last time I saw him [Derrimut] was nearly opposite the Bank of Victoria, he stopped me and said "You give me shilling, Mr. Hull". "No", I said, "I will not give you shilling. I will go and give you some bread", and he held his hand out to me and he said "Me plenty sulky with you long time ago, you plenty sulky me; no sulky now, Derrimut soon die", and then he pointed with a plaintive manner, which they can affect, to the Bank of Victoria, "All this mine, all along here Derrimut's once; no matter now, me soon tumble down". I said, "Have you no children?" and he flew into a passion immediately. "Why me have lubra? Why me have piccaniny? You have all this place, no good have children, no good have lubra, me tumble down and die very soon now".

Aboriginal voices occasionally not only decried the loss of their land to the white people but also demonstrate their incredulousness and poor opinion of non-Indigenous people's bush skills. Seweryn Korzelinski, a Polish miner at Bendigo, related how one of the Djadjawurrung women who often visited that neighbourhood

> suddenly turned her head to one side and seemed to be listening to something. After a while she jumped to a solid tree nearby and with a tomahawk split the bark and pulled out a white grub about four inches long which she ate on the spot. Asked how she knew the worm was there, she answered surprised: 'But I heard it. It was only a few steps away'.

Many goldfields writers noted what had been chronicled in an earlier period of Victorian history, that scathing humour at white people's expense was a strong element of Victorian Aboriginal culture. Oscar Comettant, a French journalist and visitor in the 1880s, perceived that 'they have that sense of the ridiculous which can be so devastating in France. They will laugh for days over some mistake they have seen committed by a white man'. Humour, Comettant observed of Aboriginal people, was a stratagem used to great effect against Englishmen who were critical and condescending towards them, adding that 'not even Voltaire himself could have replied with such droll ingenuity' as an Aboriginal man had done in response to being told '"You are an idiot … you can do none of the things we whites can": "Excuse me," replied the Aborigine, hiding a mocking smile, as well as he could, "We blacks can imitate you whites when it comes to drinking, smoking, lying, stealing, or doing nothing at all."' Gerard Krefft recorded a disparaging opinion held by Aboriginal informants on the Lower Murray River towards the white man's propensity to work unceasingly: 'There are only two things which appear great fools in their eyes, namely a white man and a working bullock'.

Lawrence Struilby's recollections include references to corroboree songs which 'mimicked the white man's ignorance of bush life, and his peculiar habits and vices', and also espied in their camps the great entertainment and humour they gained at the expense of 'any white man who was halt or lame, or in any way awkward or stupid, their mimicry of such was perfect'. The subject of non-Indigenous people breaking their promises of payment for work performed also was the subject of a corroboree song, as was a song of 'joy for the release of convicts at a squatter's establishment, on expiration of their time'.

> The Bamraman [unidentified location on the Murray River near Swan Hill] clan satirized a white man called Marsh, who employed them; but broke his word and did not pay. He always put them off by saying that the great rain (cobon walleen) had made his wagon (wheelbarrow) break down on the way. They tax him with lies (yamble), and threaten to no more wash his sheep (jumbuck) or track his horses (yarraman).

'Borrowing from the blacks'

Appropriation, adaptation or some accommodation of elements of Aboriginal material and cultural items into the dominant non-Indigenous culture was certainly occurring in the earlier frontier period. Colonists adapted to the new physical and political geography by partial synthesis and at times directly adopting rudiments of Aboriginal traditions whether it was place names, language or manufactured items, and it is evident that this continued in the gold period. Pockmarked throughout the goldfield's literature are miner's observations of how they adopted and adapted a montage of Aboriginal culture.

According to James Nisbet, a gold miner at Ballarat in 1853, the non-Indigenous miners had appropriated the Aboriginal method of communicating to one another, noting that 'Many of the diggers had learned their strange coo-ee and made the woods sing with it, as signals to their mates'. The Reverend Arthur Polehampton, a miner also at Ballarat, confirmed Nisbet's opinion of the universality of the coo-ee amongst miners, noting 'I had not been very long settled indoors when I heard a coo-ee, a peculiar call of the blacks imitated by the colonists to which I replied in like manner'. Miners such as JF Hughes (reminiscing almost 40 years after the rushes) offer some insight into why Indigenous language does not pepper non-Indigenous goldfields literature, confiding that while 'it would be interesting to learn their language [for 'amusement and instruction'], in the absence of writing materials, it was difficult to retain anything but a few words of their [Djadjawurrung?] vocabulary, such as "baan," water; "ween," fire; "narrong," bread; "na narrong," no bread'.

Bush food

It was food that provided the most potent motivation for non-Indigenous miners to explore, adapt and adopt parts of Aboriginal culture. The exorbitantly high prices of food during the initial alluvial gold rush period prompted many miners to try Aboriginal foods. The subsequent adoption of bush foods into the non-Indigenous diet led to great depletion in fauna, a subject that naturally preoccupied goldfields writers. Often miners depended on receiving tutelage in Aboriginal foods and craft. Katherine McK wrote of her childhood experiences gathering 'bushtuckers' (native cherries and wildflower nectar) commenting that in 'earlier times than ours [we] learned from the blacks what to taste and what to leave untouched in the bush wilds'. Lord Robert Cecil, who made a visit to the Kyneton diggings in 1852, recalled how the diggers at Specimen Gully 'showed me what the natives call "blackfellows sugar." It is a species of manna falling plentifully from the white gum. It tastes very much like the second layer in a wedding cake'. Some miners of course were not at all adventurous, such as JJ Bond, who was the recipient of a bush food sampling offer at Benalla in 1854:

> They are very fond of a very large grub that they discover in the rough bark of the honeysuckle tree, a Lubra brought me some one day as a rare delicacy they had been slightly roasted. I politely declined the treat and begged her to eat them for me which she did forthwith one after the other with great relish.

A substantial number of people on the goldfields, however, eagerly exploited the bush food bounty that they witnessed the Aboriginal people utilising. John Chandler, for example, noted 'great heaps of land mussel shells, which the natives had been getting out of the lagoons for years. We got some and boiled them in a bucket. They were very good with some salt'. A resident of Ballan likewise recalled that he 'often watched [Wathawurrung] lubras catching them [eels]' and also described the method of procuring the 'large and luscious white grubs [from white gums], which were a delicacy'. John Chapple and his party at the Avoca goldfields had splendid repasts of 'stewed turkey and native apples for dinner' and on another occasion '2 baskets of cockels'. One visitor to Ballarat noted the prolific amount of fauna consumed by Aboriginal [Wathawurrung] and non-Indigenous alike.

The country for many miles around on all sides was one vast forest, with
many open glades … one bird [Bustard] now very scarce in Europe are of
gigantic size and of most delicate flesh may be found in large flocks [and
are] frequently shot by the natives … miniature kangaroos abound in
the ferns but are fast disappearing in the face of civilization … a native
cat with pointed nose resembling a ferret, opossums, eels…

The repertoire of bush food included the perennial favourite duo of parrots and
cockatoos (often baked into pies), kangaroo, wallaby, wombat, ant eggs, pigeon,
parakeets, magpies, bandicoot, wattlebirds, quail, eels, native fish, dingo and
possum. Occasionally echidna, 'jackass pie' (kookaburra) and other wild fowl
were placed in the billy. James Peverell, a miner at Forrest Creek, secured a
bandicoot and considered it not 'too bad for hungry men'. James Selby, a miner
also on Djadjawurrung land, 'amused himself in the evening fishing for crayfish
and killed several possums which we consumed'.

Aboriginal people soon realised the monetary potential of bush food. William
Howitt reported that at a little distance away from the goldfields was ample
opportunity to 'enjoy the pleasures of hunting and fishing' and being plied with
bush foods by Aboriginal people. Writer Henry Giles Turner was informed by
an Aboriginal man near Benalla that he 'made his living as a fisherman, spearing
the fish, with which the [Broken] river abounds, at night and selling them to
the hotel keepers in the morning'. Howitt likewise observed that on the banks
of the Campaspe River there were: 'a number of natives fishing here, who had
caught a good quantity of the river cod, and had learned to ask a good price for
it'. This, he added with a note of annoyance, was yet 'another consequence of
the diggings'. A number of correspondents to newspapers confirmed Howitt's
observation about Aboriginal people not merely trading bush foods for trifles,
but actively striking up money transactions for the goods they sold to non-
Indigenous miners and storekeepers. Another example of such commerce can be
found in the *Inglewood Advertiser* (November 1861): 'Mr. Roff the greengrocer
had 40 brace of wild ducks, which he sold at two shillings per brace. He had
also a mallee hen and several of the eggs of that remarkable bird … They were
got from the natives, about forty miles from this place'. AB Pierce felt fortunate
that his party 'purchased from them ['a party of blacks'] a large fish of some
seven pounds, of a species which resembles the American hornpout and tastes
like an eel'.

JD Mereweather was 'asked to buy some delicate fishes, which were most
artistically arranged in leaves, and bound together with osier twigs'. He
believed that Aboriginal people belonged to an 'intelligent fine race' which
could 'calculate acutely the value of everything of which they have to dispose'.
The Murray Fishing Company, one of the largest fishing companies (established
in 1859), benefited greatly from the skills and expertise of their Aboriginal

employees. Joseph Westwood, a visitor to the Company's huts, observed how the 'fishermen reside; surrounded with a number of blacks, from infancy to old age'.

The ability of Victorian Aboriginals to exploit the market was aided by the ineptitude of many non-Indigenous miners in their quest for bush foods, such as John Chapple who 'tried night and morning for some game but could only get a teal'. Howitt also acknowledged, with some reticence, that most non-Indigenous miners and settlers lacked the necessary skills possessed by Aboriginal people to bring down the game they sought.

> The plains abound with wild turkeys; but they truly were wild, for a gathering of various tribes had lately been there, and they had been hunting them; and though Alfred and Lignum pursued them with unwearied artifice and diligence, they could not succeed in killing a single one. Emus are sometimes seen in considerable numbers; but they had fled before the natives. The ducks flew in flocks of thousands; but as there was no cover on the banks of the lake, they would not allow you to come within shot of them, and we were obliged to content ourselves with a teal and diver or two.

Others, such as Caleb Collyer, considered that 'the making of damper was a test of skill and the best I have seen made and have made was made and baked by aborigines'. JM Smith, a miner, acknowledged the superior culinary methods Aboriginal people employed when cooking possums.

> I skinned and gutted him [possum], toasted him that evening on the ashes and found him very fair feeding but rather gummy. Hunger was a good sauce and he went down slick. The aborigines do not skin them, but get some stiff clay which they carefully roll over the entire possum, then make a hole in the hot ashes and cover him up. When the clay becomes hard they break it; the skin and fur adheres to the clay and the animal comes out as clean, white and tender as a chicken, and with the above mentioned sauce, makes a good meal.

Most observers of Aboriginal culinary skills echoed the colonial writer Hubert De Castella's assessment that, whilst their cooking methods were 'very ugly to look at', they nevertheless produced 'very good eating as long as one does not have too many prejudices'. Samuel Mossman and Thomas Bannister thought very highly of Aboriginal cooking prowess, exclaiming that 'The fact is, the *chef de cuisine* at the Mansion House might add a recipe or two worth knowing to his cookery book from these natural gourmands'.

Embedded within these fusion food experiences were cultural learning experiences which encapsulated a mingling and development of what would now be described as 'cross cultural awareness'. Living together arguably influenced both Aboriginal and non-Indigenous language with new perspectives about the physical and social environment. William Dobie stated that he benefited not only from the foods they (his Djadjawurrung aides de camp) brought to his table but also the bush humour and camaraderie they freely showered upon him. Dobie considered that the Djadjawurrung 'furnished me amusement when amusement was a scarce luxury and so far as honesty and trustworthiness go, were seemingly equal at all events to their white brothers'.

Indigenising the colonists

Indigenous words such as *mia mia, willam* and *gunyah* (various spellings), denoting housing or shelter were frequently used by miners when referring to their own temporary huts. References to mia mias frequently punctuate non-Indigenous miner's documents. For example, a map depicting the Bendigo goldfields includes 'women's mia mia' and 'our mia'. Others, such as goldfields artist William Strutt, reflected in his autobiography on how he 'erected a mia mia for shelter' whilst on his journey to Ballarat. Frances Perry, a visitor to Buangor (central Victoria), in April 1852 described the familiarity which non-Indigenous people had of the name and structure of traditional Aboriginal housing:

> We took a walk amongst the wooded hills, and came upon the largest (deserted) native encampment we had ever seen. One of the Mia Mias (you know what that is by this time – the a is not sounded) … was as large as an ordinary-sized circular summer-house, and actually had rude seats all round, which is quite unusual.

One visitor to the goldfields explained the economic and functional rationale of copying an Aboriginal mode of constructing shelters at the Ballarat diggings:

> We had determined to remain in Ballarat for a few days as we could not afford speculating in deep sinking. We could not afford to remain longer looking for employment in a place where all necessities were so terribly dear. We lived in a sort of hut built of branches and bark, not unlike the mia mia of the blacks. The weather being warm and dry it was quite a sufficient shelter.

German miner and artist on the Ballarat diggings Eugene von Guérard, had been in the Antipodes less than a fortnight, but was conversant with the Indigenous name for temporary camp shelters, writing in his journal near the village of Batesford that he passed:

three or four mia-mias, the abode of some eight or ten aborigines. In front of each burned a little fire, and some spears lay at hand. The mia-mias are made of the branches of trees in the form of half an open umbrella of large dimensions. Some were covered with the skins of animals.

A significant minority of the non-Indigenous mining community, far from believing they had nothing to learn from Aboriginal people, sought to appropriate other elements of Indigenous knowledge and cultural materials from the Aboriginal people they encountered. This is a dynamic rarely explored by historians or writers examining the formation of the 'Australian legend' yet has significant implications in any discussion about the roots of Victorian goldfields society and of Australian culture. Perhaps most important was learning how to survive and thrive in the bush physically, economically and socially. Here was a degree of sub-conscious acculturation that invoked the linking of Aboriginal material and non-material culture to belonging in a new land and a new society.

Many miners expressed mixed feelings about Aboriginal people on the goldfields. This duality is found in the Faulkner family history chronicles, a tale of positive and negative memories of Aboriginal people on the Ovens goldfields: 'The Aborigines [in the Bright region], who were numerous at that time [1852-53], had not molested Ellen. William [Ellen's husband], however, had been taken by them [in a friendly sense], but become their friend after "saving" the life of one of their important men with a swig of brandy and some food'. Similarly, JG Linton recalled how, in 1854, his recently widowed mother had suffered 'for there were very few white people living in the district [Linton, central Victoria], and blacks [Wathawurrung] were numerous and could not always be trusted'. Yet, he continued, they 'were friendly to mother, who provided any sickly lubra with shelter, clothing, and medicine from the big medicine chest which she had brought with her from Scotland'. Many others, such as George Sugden who had lived in the bush for most of his life prior to the rushes and 'could understand and talk with the blacks', recalled their encounters with fondness: 'Blacks were at that time [ca 1852] plentiful and l met a lot of them [en route to Pleasant Creek Hospital]. I would talk to them and show them my [injured] hand they were kind to me'. JD Mereweather, an itinerant preacher near the Murray River, holed up by flood waters and forced to 'ensconce ourselves' amongst the sleeping Aboriginals, was the recipient of their generosity that reminded him of a New Testament parable: 'The tribe were half starved; the return of the men was looked for with impatience; this poor creature was half famished, and yet she frankly and freely offered me, a stranger, her mite – all that she had, whatever it was, and was very chagrined that l took it not'.

In the more remote districts of Victoria co-dependent relationships were very common, and benefited both peoples in caring for each other's children, mutual caring for the land, exchanging foods, cross-cultural medical advice and sharing of bush lore. These exchanges sometimes led to long-term relationships being formed and appreciation of each other's cultural perspective passed down via oral history to the present day.

In the northern states of Australia these types of pioneering co-dependant relationships have been enshrined in historical folklore, yet in Victoria they have largely remained the preserve of Aboriginal and non-Indigenous family history, and have not yet pervaded our general social and economic histories. The squatter's journals (though usually couched in paternalistic or even racist language) often contradict this monochromatic memory by giving Aboriginal people a degree of agency. A moving story of reciprocity – considered by many an integral trait of Aboriginal culture – appeared in the *Argus* (6 June 1865). It is one of the few instances I have located within the public realm which explicitly challenges stereotypical images of Aboriginal people and their passive association with the new economy heralded by the discovery of gold.

> Few colonists expect gratitude from the aborigines, but that they are not always unmindful of these obligations which go to make up what is called civilization has been proved of late in this district. Our readers will remember the paragraph which appeared in our last issue, notifying that a party of aborigines had found a thirty-ounce nugget at the Emu. This gold realized about 120 pounds for them and shortly after they had patronized the draper's shop, and provided themselves with good winter clothing, they determined to pay a visit to Clunes, where some months since a resident had been very kind to them. According to their version of the affair, he gave them money to purchase extra blankets when the weather was very cold, and they could not forget his kindness. Accordingly, the party, to the number of nine, hired for three pounds two vehicles, on Wednesday, and proceeded in them to Clunes, for the purpose of returning to their benefactor the sum he had placed at their disposal on that occasion. Some amusement was occasioned by the sight of the party when they drove out of Talbot, the women being decked in crinolines, good warm dresses, and bonnets, and the men clothed in wearing apparel of the latest fashions; but when the motive of their errand was known, they certainly rose considerably in the estimation of the bystanders.

Appraisal changes

For a minority of gold miners, close associations with Aboriginal people provided the opportunity to be schooled in elements of Aboriginal philosophy and culture and from this inculcation grew an appreciation and respect that were unusual at the time. Richmond Henty, though not a gold miner, understood the debt he owed to 'Black Charlie', being 'one of my instructors in the mysteries of Australian bush life'. A number of writers in the mining period, as in the pastoralism period, found that Aboriginal people were not as 'degraded' nor 'disgusting' as they had believed. Hubert De Castella, a visitor to the goldfields of Victoria, 'had heard so much about their ugliness that l was amazed to find them much better than l had expected' and added with equal amazement that 'their slow, relaxed gait is not without nobility, and they put their feet down with a solemnity which reminded me of the walk of actors on stage'. Robert Gow conceded in his 1861 journal (after droving with a number of Aboriginal people and forging a bush mateship relationship) that 'There are some fine traits in the characters of the blacks – they are not the wild tiger-like bloodthirsty savage generally supposed'. Miners such as JM Smith upon reflection deemed Aboriginal peoples' traditional way of life wiser than what he had at first considered:

> They are a curious race, and are said to be very low in the scale of humanity because they live without working and with very little fighting – which in my humble opinion shows their wisdom rather than their stupidity. The European makes a slave of himself for gold – and calls it industry – and then hops off the twig before he is able to enjoy it; he fights and murders his brethren, robs them of his wealth and devastates their country – and calls it honour and glory. The aborigines wander about a fine country, view the beauties of nature as they come fresh from the hand of their Maker and in their hearts they rejoice and glorify Him … They resist all his [non-Indigenous people's] attempts to make them abandon their habitual ease and independence except when tempted by rum and tobacco, for which they will readily work. It is vain to try to fetter them to houses or towns. They have tasted freedom and prefer God's canopy to man's. And for this they are called barbarians; and for this they are despised. Pshaw! The European has much to learn, although he thinks himself so very wise.

WE Stanbridge, a visitor to the central parts of Victoria, was impressed by the repertoire of information on astronomy learned through his Djadjawurrung and Djabwurrung informants, and was keen to 'produce in others the astonishment that I felt, as I sat by a little camp fire, with a few boughs for shelter, on a large plain, listening for the first time to two aboriginals, speaking of Yurree, wanjel, Larnan-kurrk, Kulkun-bulla, as they pointed to those beautiful stars'. JF Hughes had lived through both the bush frontier conflict times and the relatively quiet times of gold, and considered 'it fell to the lot of not a few who led a contemplative life and strove, Orpheus-like, to charm the wild denizens of the forest'. He had interacted with Aboriginal people in both periods and considered it a positive experience. He deemed that his intercourse with the Djadjawurrung had afforded him both 'amusement and instruction'. He admired and found great interest in many aspects of Djadjawurrung culture, and was keen to record for posterity some detail of place names, corroboree proceedings and chants, vocabulary (both pidgin English and Djadjawurrung), shelter constructions, cooking techniques, bush foods, hunting techniques, weapons and bush lore. Hughes, like many who interacted with Aboriginal people for lengthy periods in the bush, particularly enjoyed their narrating skills, sharp wit and *joie de vivre*:

> They had a keen sense of humour, and it afforded them great merriment to get me to shout aloud at night some message in their own language to their comrades across the creek, the reply which reverberated through the woods causing them intense amusement. They were also excellent mimics. One of the tribe, more adventurous than his fellows, had visited the capital of the colony, and though he ordinarily spoke in broken English he could excellently imitate the language and gesture of a new chum swell he had met at an hotel, pronouncing distinctly, with an affected air, "Waiter, bring me a glass of brandy."

Some goldfields correspondents, such as Joe Small, wrote of their slowly emerging confidence in Aboriginal people after an initial distrust had dissipated. Small's diary and poetry relates how a close relationship grew between himself and Bushby (an Aboriginal from the Ovens River region) which started off on a very rocky footing due to a difference of opinion on the ownership of a Murray codfish. A fist fight ensued and a certain degree of respect was earned by Bushby as he was considered to be 'certainly a tough customer' because of his boxing prowess. It seems that Bushby's 'being so well up in the noble art of self-defence' was one of the catalysts which enabled some of Small's racial superiority to be dispelled and a degree of respect to be grudgingly accorded. Later on in the narrative Bushby figured once again:

> I have almost forgotten to mention a visit which I received during my stay at the outstation from my black friend Bushby ['armed with a gun'] one morning, which I have no hesitation in saying was both unwelcome and unlooked for … I felt convinced that Bushby had visited the hut with the charitable intention of being revenged on me for the thrashing I had given him on the occasion when he stole my fish.

As it turned out Bushby had come to 'press me to accompany him to shoot ducks for dinner'; and 'We took shot for shot with the gun, and after an absence of two hours returned to the hut loaded with game, of which Bushby took the half'. Horatio Wheelwright, a lawyer cum naturalist, explained that when he camped at Mordialloc, 'he lived on very neighbourly terms' with the Boonwurrung people, who 'generally had their miamies close to my hut; and as I never made too free with them, or gave them a promise I did not intend to keep, I was a bit of a favourite with them'.

Frequent visits

Some non-Indigenous people reported the frequency of visits by Aboriginal people. James Morgan, a miner at Ballarat, was 'often visited by Aboriginals'. Alfred Joyce, a squatter on Djadjawurrung country in central Victoria noted 'Often in passing through the diggings township near us, l have seen them squatting about the streets or near the public house'. William McLeish, a ten-year-old boy lost and walking home on the fields of Ballarat (24 December 1856) had a solitary but moving encounter with two Wathawurrung women.

> I heard human voices in the soft musical tones of the aboriginal tongue, and almost immediately after I saw a native woman sitting at the foot of a large white gum tree – her eyes were fixed on me with a cautious searching look and I never forgot the glow that burned in those eyes, but with a kindly look in them that reassured me I walked forward and she said something I did not understand and immediately the chopping was resumed over my head, and on looking up I saw another woman engaged in chopping a possum out of a branch ... and gathering the game and blanket up, they walked away swiftly through the forest [Ballarat Common]. I saw no sign of any men or camp near at hand.

Others noted the rationale for the visits. Ray Willis of Buninyong relayed a story told to him that dates back to the 1860s of how the 'aboriginals used to come to his mother's house on the other side of the creek towards Mt. Edgerton, for fat. They used the fat not for cooking, but to rub on their bodies to keep themselves warm in winter'. Similarly the Hiscock family house in Buninyong (ca 1850s) was paid 'frequent visits by royalty in the person of "King Billy", head of a small tribe of aborigines where they were supplied with food, which they were glad to obtain'. Fear and trepidation were sometimes the initial response to visits by Aboriginal people, often followed by mutual kindness as evidenced in the experiences of a family in East Gippsland (ca 1858-59).

> When we children rushed in excitedly to tell our mother that a "whole lot" of blackfellows were coming, there was no doubt about her feelings. She quickly gathered some washing from the clothesline (fearing that a gaudy patchwork quilt would especially excite their cupidity); and then she gathered us all into the house to await developments. There were none; nothing happened. "Blackfellow sit down." They have a strong hereditary capacity for waiting. They took nothing belonging to us –

hardly a chip of wood – but they made a little fire and when eventually we had to appear, they only asked for a "big billy boil-em egg." When they were provided with our largest pot – really our washing boiler – they produced swan eggs in scores, and having boiled them, feasted and lay down; to disappear afterwards as quietly as they had come, quite pleased that all their requests for a "lil bit tchuga [sugar]," "a lil bit tea" had been gratified.

For those miners and mining town folk who did interact on a frequent or intermittent basis, it is probable that they enjoyed, and indeed, at times endured, a 'living together, living apart' relationship. Historian David Goodman also contends that the written and visual records of the 1850s contain reminders that the non-Indigenous mining fraternity and Victorian Aboriginal people 'were often in close contact'. Jack Loorham's reminiscences from the Orbost district reflect both an uncertainty and a remarkable affinity between the Kurnai and the Lohans [whites]:

I was born 23rd October, 1863, at the Station House, Orbost … There were hundreds of blacks here at that time, and a great many of them came about a few days before l was born. There were so many that my mother was scared. However she soon learned that they had come to do homage to the white child they heard had been born. When my mother was well enough to get up, she sent for them. My mother sat with me in her lap. After that they held a corroboree and the next day they went away.

John Bond, a miner at Benalla, also noted the close associations with a number of non-Indigenous people.

The natives (Blacks) are just as we see them represented. A few are now camped a little in front of this house. Benalla. There are always some in the township – women washing and so on. Men shooting ducks, stripping bark and co. for nobblers of spirit. They all are naturally of a cheerful disposition … Brandy was our favourite black man he was often in and out of the house in very free and easy fashion. All of us liked him.

Oral history corroborates Goodman's contention. Non-Indigenous family histories passed down from the mining period speak often of close relationships forged with Aboriginal identities. As a representative example, the Marsden-White family, who took up land in the Haddon-Cardigan region after their arrival on the Ballarat goldfields in 1852, has knowledge of their homestead and run's relationship to Wathawurrung sites and also to a local Wathawurrung identity called King Billy.

> The first homestead site was located on an existing aboriginal spring …
> There were a number of dairy farms based on the Bunkers Hill Ridge
> because of the aboriginal springs. Most of the little creeks draining off
> the ridge had their source at or near an aboriginal spring or soak. My
> grandfather told of aborigines in the bush on the Ridge [ca 1855]. He
> told of corroborees and he also told of aboriginal shelters and of an area
> where as a child, aboriginal bones could be found.

A number of families in the wider Ballarat district including the Comrie,
Marsden-White and Hiscock families are also the holders of rich oral histories
which speak of familial relationships with Aboriginal people. Roy Comrie's
family history has passed down oral memories of their relationship with a
Wathawurrung elder commonly called King Billy or 'Mr. Mulla, as my father
used to call King Billy'. According to Comrie family lore, Mr. Mulla often camped
at their home on the Ballarat West Common, 'taught them many ways of their
culture', visited significant sites, rode horses and sat at the family table with the
Comrie family. The bond with Mr Mulla had grown over an extended period of
time and it was considered that 'King Billy was like a part of their family when
they were growing up'. In the Newstead district too there are accounts such as
the one recorded by Thomas Martin, a child on the diggings, which describe
his family's anecdotal interactions with Djadjawurrung people, many of whom
he knew by name.

> They were great cadgers and did well cadging old clothes etc. One big
> rough old fellow with bushy hair and long whiskers used to come with
> them … He lost his old hat at the pub and my father gave him an old
> Bell-topper which he wore for months. He had a girl with him about 12
> years old. She used to do the begging. We gave her an old crinolin and
> skirt and put them on her and christened her Eliza. She thought she was
> a queen.

No doubt the presence of thousands of miners was a subject of considerable
discussion within Aboriginal society, and although we will never be able to
comprehend all the parameters, we are fortunate that we have sufficient sources
to chronicle some jigsaw pieces of the living-together story. The importance of
the synergism that occurred on the goldfields cannot be overstated. Not only
has the traditional story of gold (characterised by a mistaken assumption that
the 'Aborigines were swept aside') been shown to be untrue, but there is now
a pressing need to consider more closely evidence that the dominant colonial
culture acculturated elements of Aboriginal culture to a greater degree than has
previously been acknowledged.

7. Off the goldfields

On all appearances the first decade of the gold rush period (1850s) began disastrously for Aboriginal people in Victoria. Following on from the recommendations of the 1849 New South Wales Legislative Council's inquiry into the state of Aborigines, which called for the abolition of the Aboriginal Protectorate and offered no other coherent policy, the largely pauperised Aboriginal population, which had been shunted from their traditional lands, had little alternative other than dependent relationships with non-Indigenous pastoralists. Historian Ian Clark has pointed out that Victorian Aboriginal people's acceptance of temporary wage labour was a double edged sword, as while it afforded Aboriginal people the opportunity to reside on their natal estates, it also afforded an additional opportunity of increasing the rate of their exploitation.

Clark has pointed to the critical importance of acknowledging Aboriginal peoples' kinship system and land attachment as being a prime motivational force in how perceptions of work patterns are discerned. Some evidence for this is found in a report from Police Magistrate Andrew McCrae, to the Colonial Secretary in March 1852, on the condition of Aboriginal people in the Gippsland region. McCrae reported that all the Aboriginal people in his assigned district had served the 'full term of their employment' for a 'large' payment of money. A pastoralist on the Mornington Peninsula, McCrae understood the importance placed by Aboriginal people on employer/ employee relationships, having worked very closely with the local Boonwurrung clan resident on the pastoral station he had taken up in the 1840s.

> I would beg that it may be borne in mind that the employers, the persons mainly interested in the labour of the natives, always worked in the field with them, and saw, as it was evidently their interest to see, that the blacks had their rations according to agreement, and that they were not ill-treated by their fellow-labourers the whites.
>
> It would therefore appear … that in favourable circumstances … where the employer offers a fair remuneration, *keeps faith* with the black natives and *works with them*, that their labour, not much if at all inferior in reaping to that of the whites, may be available. [original emphasis]

McCrae's wisdom on this matter would prove to be critical for many pastoralists who were forced to employ Aboriginal people following the official discovery of gold at Clunes in 1851. Gold acted like a magnet for the vast majority of the population, including non-Indigenous sheep station workers, who left their jobs in droves. Roger Therry, a large landholder in New South Wales and

Victoria noted how in 1854 'owing to the great immigration in consequence of the gold discoveries ... we were obliged to have recourse to the Chinese and native labourers, or we should never have been able to keep our flocks together'. One pastoralist lamented: 'With every fresh gold find matters became worse for the stockowner ... In fact it became almost impossible to carry on the work of the place as more and more men went off to the diggings'. Sherer, a goldfields writer, opined that Aboriginal people's 'value, in many instances must have been incalculable at a period when nearly all white pastoral labour was suspended from the greater attraction of the goldfields'. As all hands sought to join the throng finding the 'democratic metal', it offered Aboriginal people who remained on their traditional lands, especially those whose estates were not located near auriferous fields, new prospects of increased station employment, increased wages, better working conditions and an appreciably greater estimation and admiration of their skills.

According to the figures submitted in 1852-53 by the Commissioners for Crown Lands on Aboriginal populations in the various districts across Victoria, an estimated 1,500 Aboriginal people were employed on stations and were unanimously considered to be 'of considerable service'. JF Foster wrote effusively 'many display much intelligence, and are frequently of great use to the settlers in shepherding and washing their sheep, or assisting at harvest time'. In one district it was stated 'that the whole of the tribes were employed by settlers'. One 1850s observer noted how the gold discoveries 'seem to have been highly beneficial in their operation to the Australian natives' adding that many employers, who before 'despised them', were now in a position where they had need to 'invite their services, and to deal with them on equitable and liberal terms'. Frank Shellard, a gold digger around Omeo and the Ovens Valley, described how some stations were 'entirely worked by coloured people' and added that almost all the drovers in that region were 'native blacks or halfcasts as they were all bold and daring riders and good bushmen and could pick up any stragglers [cattle] they might fall in with on their journey'. This sentiment was very evident in the 1858-59 Select Committee Report into Aboriginal people. The majority of the respondents from each area of Victoria duly noted the prevalence of good wages being demanded by Aboriginal people.

During the early phases of alluvial gold mining in Victoria, then, Aboriginal people assumed an importance as a labour force which had not been seen since the first period of colonisation almost 20 years prior. From across Victoria came reports of their vital significance as a work force, which enabled the pastoralists to continue in the face of a labour shortage. 'It is a fact I should like to state, well known to me', wrote former Assistant Protector of Aborigines in Victoria, Edward Stone Parker, 'that, at the time when the country was in a state of universal excitement on the outbreak of the gold mining, there were several stations where no shepherds were left but aboriginal shepherds'. AC Cameron, at 'Terinallum' in the western district of Victoria, wrote in a series of letters between 1851-58 of the severe labour shortage crippling the pastoral industry, only relieved by Aboriginal workers; orders drawn in November 1854 demonstrate that Aboriginal people made up perhaps 75 per cent of Terinallum Station's workforce.

> [F]or it is all Gold in this neighbourhood – every body totally Ignorant about wool … the shearers are doing their work moderately well, and I have to be pretty civil; the most of my shed men are my faithful darkies … William, Cocky, Jamie, Billy Downie, Charlie … I have for sometime been shepherding one flock myself, and have another to be thrown on my hands on Monday; but if two of the Blacks that have promised come tomorrow I will be able to jog on a few weeks longer without doubling the flocks … do you think you could get me any Bullock drivers for the wool? I see no chance of getting any here … I wish you would give my Blackfellows a hint if you see them about the Leigh [River] … I am washing the sheep now. I have got a lot of Blacks engaged for washing, they are doing very well as yet; I have them all under a written agreement for Six Shillings per week.

Aboriginal workers were deemed to be highly proficient in a range of skills including: bark stripping, washing and tracking sheep, bullock drivers, shearers, fire fighters, as guides across unfamiliar country, general hands, wool pressers, scourers, rouses, carters, musterers, timber cutters and fencers. The full employment of Aboriginal people during peak times in the pastoral industry is evidenced by a letter received from the Yelta Aboriginal Station advising that 'we got no blacks they are all gone shearing'. Newspaper correspondents from remote areas in Victoria attested to the fact that 'blacks were employed and they turned out their shorn sheep in a far more satisfactory manner than the majority of the whites so engaged'. At various stations it was noted that Aboriginal workers were sometimes paid the same rate as non-Indigenous workers. William Moodie noted 'I always paid any blackfellows working for me … The men insisted on being paid for their wood-cutting in white money [gold] – and got it'.

Rural workers accounts such as George Sugden's reminiscences of pioneering life in Victoria are peppered with references to over-employed Aboriginal people being the mainstay of station life whose tasks included providing an essential variety of fresh foods, cooking and being servants, providing sexual services, performing the onerous task of sheep washing, mustering in areas that were dangerously inhospitable, tracking lost non-Indigenous workers and retrieving valuable errant stock. Living on a remote Victorian station, Sugden was 'quite in his [Aboriginal aide's] hands and knew that as long as I stuck to him I was safe'.

He acknowledged, without reservation, the vast superiority of his Aboriginal co-worker and came to the somewhat disconcerting, yet pragmatic conclusion that work and survival in the bush was a great race and class leveller.

> I was given a half caste named Davis to help us. He was a splendid stockman, none better. We lived together and slept in the one tent … Though I did not like cooking for blacks … [Sugden reluctantly ended up cooking for Davis as] I saw it was the best way … I never saw a man use a whip like Davis. Within a week Davis had the horses so trained that they would come up to the tent and stand till I had picked out the required horses and then move quietly away to feed.

Though non-Indigenous bush workers and pastoralists clearly perceived Victorian Aboriginal bush workers as the finest workers from a skill and knowledge base, there was a barrier to their entry into the non-Indigenous pantheon of bush worker mythology that Russell Ward wrote of in his seminal work *The Australian Legend*. The barrier was the Aboriginal bush workers' steadfast refusal to behave like English country workers under the auspices of a genial pastoral overlord. Historian Richard Broome persuasively argues that Ward 'celebrates the roving, independent stance of rural workers as the seed bed of Australian egalitarianism, but similar Aboriginal behaviour is never associated with any nationalist mythology of worker independence, as it might well be'.

NA Fenwick, Crown Commissioner of Lands in 1852, commented on the Aboriginal bush workers with a deal of frustration, 'they will work only when they choose', a sentiment echoed by other Crown Land Commissioners in Victoria. Robert Gow, though waxing effusively about Aboriginal workers' legendary bush-work acumen, was typical in adding (with some exasperation): 'But no gifts, no kindness, no education, no comforts, nothing can win them from the chasm of their wild nomadic life'. Gow and many others failed to discern Aboriginal people actively choosing to be two-way people – that is, opting to balance their traditional lifestyles with some participation in the non-Indigenous economic milieu. Other observers of Aboriginal bush workers understood there were legal and cultural imperatives motivating their work patterns. Some pastoralists, through their long associations with Aboriginal people, such as Peter Beveridge near Swan Hill, understood and accommodated such work habits realising to some degree that a great number of Aboriginal people were neither dependent on non-Indigenous people's material goods nor willing to forgo their Aboriginal cultural imperatives for a life of dreary servitude. W Dobie, a farmer on the Richardson River in central Victoria related how 'Doctor Syntax', his Djadjawurrung farmhand, was extremely fond of the money he earned but Dobie acceded to the knowledge that Djadjawurrung political and social obligations were far more pressing than weeding a white man's vegetable patch.

The prospect of money was the Doctor's most convincing argument for digging up the weeds. "He must have all done by a certain big one Sunday," he would tell me, "as Wimmera black fellow would come *barley* then, and then black fellow all pull away along Avoca, and then big one corrobary." I admitted the propriety of this, and urged him to go a-head.

8. Social and environmental change

Notwithstanding the evidence of cohabitation documented in chapter seven, the goldfields were an inherently violent and dangerous landscape, especially for Aboriginal people. Acts of violation such as sexual abuse, poisoning and shooting of dogs, desecration of graves and interference in Aboriginal affairs were frequent. Whilst the majority of commentators enthusiastically extolled the Victorian goldfields as relatively free of the violence common on the Californian goldfields, historians such as David Goodman have peered into the historical records and now insist upon an 'edgier interpretation' of Victoria's goldfields.

An examination of newspaper reports and court records of the early gold rush period bears out Goodman's argument. By the 1850s, several generations of Australian colonial society had either lived alongside the brutalised convict class or were closely aligned with them as family, friends or fellow workers. Testimony of how 'nasty, brutish and short' life at the goldfields could be is underscored by how inured miners and other commentators became to what would now be termed horrific murders and, equally, to industrial accidents. One miner expressed his dismay at the goldfields violence thus: 'Society is in an awful state at these diggings; four murders within the last month. On the night before l left the diggings a man was shot at in an adjoining tent'. Others were more nonchalant about the deaths and accidents.

The records of violence and murder against Aboriginal people by non-Indigenous people during the gold rush period are very extensive. PC Chauncy reported two extremely violent sexual assaults by non-Indigenous miners upon Aboriginal women at Whroo and Rushworth diggings.

> A fortnight ago, when I was at Rushworth, a white man had just been sentenced to 3 months imprisonment for abusing an *old* native woman. She had been made helplessly drunk and then five men abused her so as to nearly cause her death; she could scarcely crawl into court the next day; this is no uncommon case. I was informed that a native woman lately died at Whroo, from the abuse she received from a number of white men.

Retribution for minor incidents by Aborigines was swift. Surveyor, John Wilkinson recalled:

> Here [Nicholson River crossing near present day Bairnsdale] in 1856 Charles Marshall erected a hotel and store ... The store was popular with the diggers but it met a sad end. Two Aborigines stole groceries and

spirits from the wattle and daub building attached to the hotel. Marshall caught them and shot them, and in retaliation, a number of Aborigines burnt the store down, so that it had to be rebuilt.

A report in the *Ararat Advertiser* entitled 'An affray with the Aborigines at Cathcart' (10 August 1858) demonstrates that inter-cultural relations in the hub of goldfields towns were often volatile. The newspaper's account implies some revengeful attack being enacted on the Ballarat Medical Hall by Djabwurrung people and a subsequent quarrel with a Police Inspector. The reporter contended that 'but for the timely arrival of [Police Inspector] Mr. Smith that some serious damage would have been done'. In 1862, Buckley, an Aboriginal man, was violently murdered by two miners at Mia Mia Flat. Numerous newspaper reports revealed that Buckley was a 'familiar figure' to the diggers. The *Mount Alexander Mail* provided a précised account of the murder:

> An aborigine named Sampson found Buckley in a waterhole – inquest conducted "The general impression as to how the murder was committed appears to be that on Saturday night Buckley proceeded to Simm's tent to persuade the lubras to return to their encampment; that his importunities to this effect excited the wrath of Simms, who struck him several times with a piece of wood taken from his stretcher, and then pushed him outside to the waterhole. Such at all events is the substance of a statement made by an aboriginal boy who alleges that he slept the night in Simm's tent, and saw all that occurred. Of the two lubras we have not yet heard anything, nor do we know whether they will be called upon to give evidence.

Occasional newspaper reports of thefts or assaults by Aboriginal people upon non-Indigenous people were at times reported with a comical tinge, explicitly implying Aboriginal peoples' harmless 'child' like propensities. A report in the *Argus* exemplifies this genre of reporting about Aboriginal people:

> On Monday night three of these children of Ham "stuck up" a resident in Sale in the street, and would not part hold of his garments until he had given them "white money" … shortly after a police constable appeared and took the aborigine in charge. The constable, however, not being thoroughly up to the subtlety of the predatory tribes, took hold of the offender by the blanket, and proposed to give him a night in the lock up. His sable prisoner, however, thought different, and quietly withdrawing the pin which held his blanket, left the garment in the hands of the guardian of the peace, whilst he slipped quietly away.

Encounters of the nature of an organised, violent or group attack however, were not widely reported in the gold period, though reports such as one that appeared in the *Argus* (20 September 1855) amply demonstrate some Aboriginal people's determination to independently enact their laws and if need be use their weapons upon any non-Indigenous people who would seek to divert them from the performance of their lawful duties:

Extract from letter of Surveyor in charge of Gippsland District:-

"Since my last report I have seen other individuals from the Omeo goldfields who corroborate the statements made therein.

"A large party of blacks, numbering between one and two hundred, have come down from the upper part of this and the neighbouring districts to retaliate the onslaught made on the Swan reach Tribe some time back … and came in a strong body to the neighbourhood of Sale. They named five aboriginals whom they expressed their determination to murder, and intimated to the authorities that if interfered with by the police they would fire on them … I myself unexpectedly came upon their camp one night, and had a "yabber" with them. I have seldom seen a finer body of blacks. They robbed one shepherd's hut, but aver their intention not to molest the 'whites,' unless interfered with."

Stories were often told by miners of 'murderous attacks made on early squatters' by blacks, but generally the consensus from official commentators was 'The worst that can be charged against the blacks was a predilection for stripping men's shirts and pants off the clotheslines'. There were, however, many on the goldfields whose irrational fears of encountering 'wild' or 'myall' Aboriginal people may have precipitated inter-racial violence. Henry Boyle's memories of coming to Buninyong in central Victoria just prior to the gold rush are perhaps typical of many non-Indigenous people's strongly held, but ill-conceived, perceptions: 'we passed many black camps, being very careful to follow Mr Jamieson's (of Buninyong) advice:- "Do not let a blackfellow follow you; but always keep him ahead of you." They were very treacherous, and would often split a white skull open with a tomahawk without any apparent provocation'. Mistrust and fear of Aboriginal people peppers non-Indigenous family narratives, as can be found in many local history publications.

Reminiscences of the goldfields contain accounts of great trepidation about Aboriginal people, such as in the diary of Thomas Booth: 'The Aboriginals were quite numerous around Buninyong [central Victoria] … When the tribe was seen approaching we retired to the inside of the house and remained there until they went past'. Oral history from the Coxall family, also of Buninyong, mirrors Booth's: "Uncle Tom used to tell us a story of a tribe of Aborigines that used to come from Burrumbah [Burrumbeet?] for their corroboree, an assembly of sacred festive, or warlike character, every year, he was a boy of 3 years old and all the kids were frightened of them. There were thirty-six of them, with their own King Billy."

Alcohol and substance abuse

It was commonly alleged that alcohol abuse was the reason for inter and intra-racial troubles during the pastoral period. In the gold rush period the further breakdown of traditional law increased the level of abuse, and hence the level of conflict. TH Goodwin, a missionary on the Lower Murray argued that:

> Many of their best customs and most stringent rules, in regard to the young people, have been weakened and broken by the introduction of the evil habits of vicious white men; and the young men being more intelligent, pay less regard to the old men, and follow their own sexual desires. The young women are even more sensuous, and reckless of future consequences.

Aboriginal informants in court cases and other historical sources testified to the immense internal conflict that occurred within their communities which they attributed squarely at the feet of alcohol abuse. Visitors and travellers to the goldfields such as James Bonwick, were unequivocal that alcohol abuse was exclusively the reason for the demise of Aboriginal people, exclaiming: 'It is not the want of food, nor is it mere disease, that occasions the evil; the sorrow, the demon, the destroyer is *Strong Drink*, under whose maddening influence murders are committed, and fatal conflicts induced'.

But it was not just Aboriginal people who abused alcohol, as the following portrayal of shared alcohol abuse reveals: 'I found John Cookey drunk and bleeding from fighting, John Weir bleeding from fighting, John Wearing cut and bleeding from fighting John Weir, Teddy our blackfellow drunk … John Cookey was so overpowered by liquor that l had to give him large quantities of ammonia to get him sufficiently recovered'. Alcohol abuse and the often deadly immediate effects of an unregulated alcohol industry were widely reported in the colonial press. Samuel Lazarus met a man whose 20-year-old brother had died from the effects of drunkenness at Bendigo: 'It is painful to contemplate the horrible havoc which drunkenness makes on the diggings'. Visitors to the Victorian goldfields were painfully aware that alcohol abuse was a terrible scourge upon non-Indigenous and Aboriginal people alike. Some, such as James Bonwick, felt a sense of responsibility for Aboriginal people's alcohol abuse:

> The miserable remnant I saw suffering with the Whites, from the effects of wild intemperance; it being then Race week, an awful time of reckless

extravagance and unbridled debauching. The shrieks of drunken women, the cries of reeling natives, and the quarrels of besotted men greeted me upon my first visit to Casterton ... Yet the law pretends to protect the native against him who holds the bottle to him; but who is to enforce the law, or who does enforce it?

Excessive alcohol consumption was observed to impact on Victorian Aboriginal people's mobility, motivation to carry out traditional food gathering practices and ceremonies, and liability to mining camp accidents such as falling down shafts. Objections to and schemes for the abeyance of excessive alcohol consumption were frequently sported in colonial newspapers of the period and by the respondents to the Victorian Government's circular questionnaire on Aborigines in 1858. One correspondent in the *Argus* called for compensation in place of 'maddening poison in the shape of fiery drink' and posited that 'the law, if it will not prevent the white man from brutalizing himself [with alcohol], should prevent him from enabling the ignorant native to become equally brutal'. Miners and correspondents in colonial newspapers such as the *Argus* repeatedly pointed to the 'great annoyance' caused by Aboriginal people who had been plied with alcohol. The cause of their ire was said to be Aboriginal people's 'objectionable habits; their continual drunkenness and the noises they keep up at their camp at the wharf [Echuca] all day and all night'. One observer noted

> The blacks are almost all fond of intoxicating drinks ... It is hardly possible to prevent them from obtaining drink, as they have as good a right to spend what they earn as a white man. Their young men get a pound or two occasionally by cutting bark, tailing cattle, and c., this they almost always lay out in drink, and treat all hands at the camp. They are not at all selfish amongst themselves, but they are so as regards the whites.

The gold rush and the concomitant gold towns and diggings shanties were considered to be the harbingers of destruction for Aboriginal people in Victoria, and according to Alfred Joyce, a pastoralist in central Victoria, effectively sounded their death knell. Joyce wrote:

> The blacks did not show any signs of serious diminution till the breaking out of the diggings, but their demoralization had been going on all the time previously. Debauchery and drink was doing its work. When bush inns became numerous the blacks congregated about them and took all the drink that was offered them, and purchased it whenever they could get a coin or two by begging or otherwise. All this was bad enough when the white inhabitants were few and far between, but at the outbreak of the diggings, with greater temptations and facilities, swept them off rapidly.

Indigenous temperance

There are, however, a great number of reports which relate to Aboriginal habits of temperance. Notable examples of abstinence were recorded by goldfield newspapers correspondents such as one from the *Daylesford Mercury* which remarked (6 May 1865) that a small party of Aborigines of the Daisy-Hill tribe (presumably Djadjawurrung) had discovered a nugget weighing two ounces on Amherst Flat, and being 'Wiser than some of their generation, instead of spending the money realized by the sale of the nugget on drink, they purchased a stock of warm winter clothing, and with this on their backs they paraded the streets with the greatest possible dignity'. A month later a correspondent for the *Talbot Leader* informed his readers of another discovery of a large (30 ounces) gold nugget by Aboriginal people at the Emu goldfield which realised about 120 pounds for them. The successful party hired two vehicles to take them to Clunes, with the purpose of fulfilling their kinship obligations to a non-Indigenous storekeeper there. The newspaper correspondent and other observers keenly noted their abstinence from alcohol, stating:

> One thing was evident, namely, that they were all quite sober; and on enquiring how this was, since aborigines are expected to get drunk the moment they obtain any money, it transpired that several attempts had been made in certain 'shanties' to induce them to drink, and that they had refused point blank to imbibe anything stronger than ginger beer. Indeed one of the party, who appears to be head or chief of the rest, replied to one of the tempters that 'black fellow could be a gentleman as well as whitefellow,' – meaning, we presume, that he was not bound to get drunk because he had suddenly acquired a considerable sum of money. In this respect this sensible aborigine is decidedly in advance of some of the whitefellows, to whom the sudden acquisition of fortune is more often the prelude to intoxicated habits than the forerunner of staid and sober conduct.

James Sinclair encountered two Aboriginal men in his travels across Victoria who 'both expressed what is a rare thing for a blackfellow to do, their thorough disgust of drunkenness and drinking, which caused, as the eldest black rather expressively stated it, both white fellow and black to become "too much ----- big fool"'. Reports from the honorary correspondents and William Thomas, Guardian of the Aborigines also noted Aboriginal leaders on the goldfields such as Djadjawurrung elder Biebie (Eliza) who exhorted her countrymen and women to 'leave of that beastly drink'. Some 'steady' Wathawurrung elders too chided their people by frequently telling them 'blackfellow soon all gone if drink much grog'. Barry Collett, in his history of South Gippsland, wrote of initiatives by

Bratauolong elders to stem the destructive effects of alcohol abuse amongst his community, which, sadly, non-Indigenous authorities proved unable (and unwilling) to administer. Collett wrote:

> In October 1853 Old Darby, a Bratauolong elder, possibly from the Corner Inlet Kut-wut, asked Police Magistrate Tyers to order a Sale publican not to sell more than one glass of grog to Aborigines, for he and other elders were worried about the way in which Bratauolong youths increasingly neglected their traditional responsibilities.

Kulin elders at Coranderrk, an Aboriginal reserve established in 1863, had also created their own effective stratagems to manage excessive alcohol consumption such as creating their own reserve court which set monetary punishments for drinking and the threat of forfeiting the right to marry.

Laws to prohibit the supply of alcohol to Aboriginal people were enacted and enforced with great regularity by the Victorian police across the Colony, and it was claimed that 'numerous magistrates throughout the colony have expressed their intention to refuse to grant a license to any publican against whom this offence may be proved'. But it proved impossible to stem the sale of alcohol. One report in the *Argus* expressed the frustration of the police and magistrates in Victoria who were 'exerting themselves to put down the sale of intoxicating liquors to the aborigines, and are dealing with such cases summarily, there is not a similar activity on the New South Wales side'.

Cultural interference

On the Victorian goldfields a number of notable instances occurred where non-Indigenous people interfered in Aboriginal cultural affairs. Perhaps the most potent reason why very little discussion has taken place is the derogatory and often erroneous manner in which such interferences were reported. In many gold mining towns, non-Indigenous people witnessed Aboriginal people practicing their customs and laws, and these were generally reported with an air of incomprehensibility and naiveté. The faux pas connected to mortuary ceremonies was particularly disturbing to Aboriginal people's sense of propriety and decency as they witnessed non-Indigenous miners ploughing up the land and desecrating mortuary sites. The miner JC Hamilton noted:

> I was at Bendigo, in the year 1854, at the diggings, and, arriving late one evening, our party drew up near a log, which we intended to light for our fire, but discovering a newly made grave just behind the log, we lit our fire away from it, and used the logs for a seat. A party of blacks [Djadjawurrung] came and wanted us to shift from the place, as they had

buried one of their number there late the evening before, and wanted to complete their arrangements. We told them we would not interfere in any way, so they set to work and put up a brush fence round the grave at a distance of about twenty yards, leaving an entrance at the furtherest point.

Not all miners were as respectful as Hamilton and his mates. Korzelinski, a miner at Sydney Flat, near Bendigo, whilst digging a shaft came across the grave of a 'native [Djadjawurrung] chief' and considered it would have been 'an excellent find for an archaeologist'. He offered that his only reason for not interfering with the fresh grave site was 'I was too busy sinking my shaft to worry overmuch about scientific problems'. Others such as A Batey in 1856 resolved to 'dig for an aboriginals cobbera (head)' and rifled through an Aboriginal graveyard to achieve his goal.

On the goldfields itself, miners such as Carl Lagergren claimed that he and others had often acted as mediators between Djadjawurrung men and women when alcohol abuse occurred in their community:

When they have been fortunate enough to get hold of spirits … it does not take long before they become trance like and start a so called "corroboree" or dance where both men and women start to make a lot of noise and douse ear piercing noise … A truthfully scary scene which becomes even more scarier when the men reaches a more trance like behaviour and starts to hit the women with all their powers and these women see themselves as fortunate if there was nearby a tent occupied by white people. This way she could seek protection against the mens loving hand, showing their love … and many times it happened to me as well as my friends that you had to go out and perform a piece [peace] making act.

A Batey recalled that a party of non-Indigenous people's appearance at the scene of 'an aboriginal beating his lubra' put a 'stop to the proceedings', whereupon the Aboriginal woman cried 'That one always him beat me'. From gold mining areas where internecine fighting and feuding still raged, such as in Gippsland, there were reports of non-Indigenous people and authorities attempting to act as peace brokers or as hosts for those seeking refuge.

William Craig, a miner at Mount Cole, described how a party of non-Indigenous miners formed a 'council' to deliberate on what to do about three Djabwurrung men who they believed had committed a payback punishment on a neighbouring clansman. Though lynching them was seriously contemplated it was considered the best course was to send for a local justice of the peace. A messenger was dispatched, but the magistrate had no police to assist him and was too busy to attend. Craig duly noted the slender regard afforded internecine legal matters by the authorities: 'As evidence of the slight importance attached to the murder of a native at that time, it is worth noting that, although information of the crime was later on sent to the police, no one ever appeared to make official enquiry regarding it'. There were some occasions when non-Indigenous interference into customary feuds or cultural issues was initiated by Aboriginal people. Korzelinski referred to an instance when feuding Aboriginal groups were to meet together on a friendly basis and that 'To stress the solemnity and importance of the occasion, English neighbours from a nearby settlement were also invited', presumably to perform a role as peace observers.

Meddling in marriage

At times Aboriginal people demonstrated a great deal of resentment and retaliation towards non-Indigenous people for meddling in Aboriginal marriage customs. William Thomas, Aboriginal Guardian, reported in 1851 that a group of Aboriginal people (Djadjawurrung) had surrounded his cottage in Melbourne and had sought out and demanded a 'poor lubra' named Polly, who had sought refuge with Mrs Thomas. It seems that Polly had lived with a non-Indigenous shepherd on JM Sanger's Avoca run and was considered to be 'one of the most modest, well-conducted females, in many respects highly civilized'. While the Guardian's wife hid Polly in the dining-room, Thomas persuaded the angry group of Aboriginal people to leave. Later, Thomas reported that Polly reconciled with her shepherd, married him and returned to Avoca. Another instance was reported by Reverend Mereweather interceding on behalf of an Aboriginal woman named Lucy, of an unidentified 'Murray River tribe', who had been betrothed to Charley, a 'black fellow'. Charley came to Mereweather 'humbly petitioning that I would persuade [Lucy] to give herself up as his wife'. Charley's imploring fell on deaf ears, much to his disgust. A very similar event appears to have transpired at Anderson's Creek Diggings in 1860. A non-Indigenous miner provided evidence which

> showed that the blacks [Kurnai clansmen: Tarra Bobby and Billy Logan] had given great provocation to a white woman who had a black lubra in her hut; and that the man, to frighten the blacks away, took up a gun, and when the blacks were about a hundred yards off fired, not knowing it to be loaded … The white man was fined 5 pounds, as damages to the two blacks.

Environmental degradation

The environmental changes that occurred as a result of both alluvial and deep lead mining were profound. This was the second wave of catastrophic environmental degradation to impact on Victoria since European colonisation. The first phase was the pastoral period, beginning effectively in 1835. The landscape that miners viewed was not, therefore, a pristine one, though it was often portrayed this way in their writings and artwork. They seldom acknowledged that they were, in fact, part of the second wave of dispossession and environmental devastation.

The pastoralists were in no doubt as to the disaster that millions of sheep had wrought upon both the environment and Aboriginal people. The extremely rapid destruction by sheep of plant foods that Aboriginal people had previously subsisted on manifestly altered the ecosystem and had flow on effects on inter-cultural relations. Aboriginal informants complained bitterly to Victorian squatters and others throughout the 1830s and 1840s of the profound changes to vital plant ecology caused by the introduced livestock.

To the miners, however, pastoralism and Western-style agriculture were largely viewed as benign. A common theme in miners and other commentator's writings and pictorial works is that mining alone was the great despoiler of nature. One result of this was an inability to understand that even in areas where mining had not utterly changed the physical landscape, the essential bio-mass had already been irrevocably transformed by pastoralism. Alluvial gold mining, historian Barry McGowan points out, even in its more primitive forms, 'affected the environment from the outset and was not as benign in its effects as present day landscapes might suggest'. Some goldfields correspondents were in no doubt as to destructiveness of the miners' presence. 'The diggers', William Howitt observed, 'seem to have two especial propensities, those of firing guns and felling trees ... Every tree is felled, ... every feature of Nature is annihilated'. Thomas Woolner considered that the Mount Alexander diggings in 1852-53 was 'what one might suppose the earth would appear after the day of judgment has emptied all the graves', whilst Louise Meredith considered the mining landscape to be 'more irredeemably hideous than the bleakest mining village in any English coal or iron district'.

The Central Board of Aborigines alerted the government to the 'vast quantities of fish destroyed annually by netting and the swivel gun ... Both fish and game are ruthlessly killed in such a manner as to injure, not only the interests of the blacks but those of the colonists generally'. The declining fish stock in Victoria was so serious a matter that specific legislation (Fisheries Bill) was drafted in 1873. Acknowledgement of the catastrophe that had befallen Aboriginal people in particular, whose traditional and adopted monetarist economy depended very heavily on the availability of good fish stocks, can be gauged by an amendment proposed to the new bill, which was accepted without discussion, 'having for its object the allowing Aboriginal natives to take fish in any way for their own use'.

Even more serious was the damage to the environment and the Aboriginal cultural landscape done by gold mining practices such as hydraulic sluicing and later dredging; eventually regulations came into force which sought to limit the obliteration to water courses and adjacent land. The immense damage to the physical environment caused by gold mining operations was commonly said to be the reason for the absence of Aboriginal people in a region. Typical of this appraisal is Michael Goonan's, whose 'mother's people lived, in 1854, not far from Yackandandah'. Goonan contended that the tribe in this area were 'shy and quite harmless' and that 'By degrees they shifted back to more uncivilized parts until a black fellow was rarely seen'. Similarly, Katherine McK noted that Aboriginal people had caught 'bandicoots, wattlebirds and blackfish from the creek' near the Daylesford diggings until 'men with guns and dogs came from the mining towns and camps, and soon the wild game was exterminated, even on the rough ranges'. In the central Victorian town of Scarsdale it was noted that the 'blacks [Wathawurrung] came in to the diggings looking for water'. A number of correspondents observed that 'their original food is getting more and more scanty' and thus obliquely deduced the connection between their attachment to the goldfields where money and food was available and their shunning of the degraded forests and plains where subsistence was problematic. Undoubtedly, the destruction of the natural environment by gold mining did indeed prompt the 'withdrawal' of Aboriginal people as camping and hunting places were despoiled, yet there is evidence that the breaking up of large pastoral leases also led to environmental degradation which also profoundly impacted on Aboriginal people.

Not all areas suffered equally; for the intensive alluvial goldfields were spread out, both geographically and chronologically. Aboriginal people were reported, by almost all the respondents to the 1858 Select Committee's enquiries, to be eking out an existence amongst the mining towns or residing on pastoral runs. Dislocation from their lands and cultural heritage produced stressed communities who engaged in the internecine violence that observers on the goldfields noted with regularity. But, as time passed, Goodman argues, 'less would be heard of the injustice of a people deprived of their traditional sources of food, and more of the "propensity of the race" to this or that weakness'.

The significance of less visible mechanisms by which Aboriginal people in Victoria were dislocated, such as interference in traditional mortuary ceremonies, food gathering practices and legal matters, cannot be understated. The altering of traditional living patterns was a contributing factor to a community wracked by alcohol abuse and alternately neglected and controlled by colonial governments and Christian missionaries.

9. Governments and missions

The 1850s have generally been described as a decade wherein Aboriginal people were overlooked by the new Victorian Government (separation from New South Wales occurred in 1850). Judging from the almost universally negative response about the level of government assistance afforded to Victorian Aboriginal people during the 1850s it is hard not to agree. In answer to the question (asked by the 1858-59 Victorian Select Committee of the Legislative Council on Aborigines) 'Has assistance in the form of clothing, food or medical attendance, been bestowed on the aborigines of your district by the Government' the answer was a resounding 'No assistance of any kind has been given them by the Government'. This fact is borne out by the amount of expenditure for the Aborigines of the Colony of Victoria in the seven year period from 1851-58; 12,000 pounds, of which over half was expended on the salaries of non-Indigenous people, and only a little over ten pounds was expended on 'Medicines and Medical Attendance'. Some writers, such as EJ Foxcroft almost a century later, considered that 'Euthanasia in fact, has been the aim of native policy in Victoria after 1850 ... The policy adopted between 1850 and 1860 was, as can be seen, a half hearted one'. More recently, historians such as Ian Clark and Michael Christie have argued that 'This decade may be characterized as one of Government neglect of the Aborigines'.

Many commentators in the 1850s were highly critical of government's treatment of Aboriginal people. 'You have taken possession of a country that is not yours', RL Milne insisted, 'Ye have disinherited and slain its owners'. A lengthy editorial in the *Argus* (17 March 1856), frequently punctuated with 'bitter indignation' at the government's 'basest meanness and dishonesty in our treatment of this unhappy race', insisted that the 'wretched pittance' allocated to Aboriginal affairs be replaced 'for fair and even liberal treatment'. The editor was adamant that no expense should be spared and contended that:

> We mean what we say, literally. We would feed and clothe every black in Victoria, and would do this regardless of expense. If it cost ten thousand – well! If twenty thousand – well! If a hundred thousand – still well! Were they able to strike a bargain for the land, we should gladly purchase it at hundreds of thousands of pounds. It is dishonest to withhold it, because they are ignorant and helpless. We would feed them and clothe them as long as a black was left amongst us, and when the last was gathered to that Creator of whom he at present knows so little, we should rejoice to think that at the last great day, he could not arraign us for having behaved towards him here below, like a tyrant, a coward, and a swindler.

There were, however, also correspondents who considered that institutions established for Aboriginal people such as the school and refuge at Mt Franklin in central Victoria were a 'complete and sufficiently transparent failure', with the implicit judgment that it should be abolished. Questions were asked in Parliament about why the Goulburn and Loddon Aboriginal stations, given their proximity to the gold diggings, were not 'parcelled out and put up to auction in suitable lots, for the benefit of the revenue and encouragement of an agricultural class on that part of this very rich gold country'.

Increasingly, any discussion during the 1850s in relation to supplying Aboriginal people's physical and spiritual needs (for the two were usually viewed as inseparable), polarised into two camps or viewpoints. Some writers believed that as they were dying out and their 'barbaric tendencies' had not diminished, nothing should be done, bar providing for their immediate physical wants and recording as much as possible about this 'interesting branch of the human family' before they became extinct. Most favoured saving the youths by supporting the philanthropic works of concerned individuals, rather than government initiatives involving capital expenditure.

Honorary correspondents in the goldfields regions such as Andrew Porteous at Carngham, near Ballarat, certainly ascribed to this view, and considered both reactive and proactive measures were necessary to stem the abuse of alcohol consumption among Aboriginal people near the goldfields and towns:

> A few of the young men are generally employed on stations, and receive a small remuneration, but all they receive, both for labour and opossum rugs, is spent on intoxicating liquors, and I fear they will not leave off this evil habit unless prohibited from visiting the goldfields and are allowed to settle on some portion of land where they would take an interest in improving it.

The same issue had been expressed during the Aboriginal Protectorate period (1838-1850), but reached its zenith during the gold rush. As already explained, gold fossicking, trade in possum skin rugs, baskets, primary produce, and employment on pastoral stations after 1850 afforded Aboriginal people a new degree of economic independence. The damaging social effects of alcohol abuse and the absence of paternal control were a concern re-iterated many times by well-intentioned Correspondents and Guardians. In his June 1871 report, Porteous advocated a pass system, as he found the local Wathawurrung people could not be restricted and regulated sufficiently to keep them from their commercial activities in the goldfields and towns:

> The tribe still follow their occupations of fishing, hunting and making of opossum rugs, which they barter for stores, but often for grog. It is

almost impossible to keep them from visiting the towns, and yet they have no business to transact in those towns except begging for grog and making themselves liable to be arrested under the Vagrant Act. They have no hunting field nor fishing river within these towns, and if they have anything to sell let them apply to the local guardian for a pass for that day, to be within a town to be named in that pass. Most of the tribe are old and feeble and unable to do any work. The young men are able and willing to work, and some of them can do work as well as any white man, but they are like any of the white men, and would spend every shilling they earn upon grog, if they can possibly get it done.

The frustration of Honorary Correspondents such as Porteous and others to restrict Aboriginal people from frequenting the goldfields, readily obtaining money and over indulging in alcohol, combined with the Victorian Government's persistent refusals to fund Aboriginal welfare, beyond providing for food, clothing and shelter, greatly contributed to a philosophy of centralising Aboriginal people onto a few reserves. Persistent calls for dictatorial control over Aboriginal people as the only feasible means of caring for them were also heard. In 1863 the Central Board for Aborigines (CBA) identified in excess of 120 Aboriginal people outside of the Honorary Correspondent's influence. In addition it was reported that some Aboriginal people in the central Victorian goldfields were not included in the Board's population estimates. In the CBA's fourth report (1864) the matter of forcibly removing Aboriginal children was officially broached, principally those who lived in proximity to the goldfields: 'it is well known that the blacks were in the habit of visiting the towns and goldfields very frequently, where they readily procured intoxicating liquors. They were ill-clad and ill fed; their children were uncared for'. The CBA made it abundantly clear that they sought much greater power to intervene in Aboriginal peoples' lives by making 'urgent solicitations for some amendment of the laws affecting the blacks'. It plied the government to introduce Bills for the 'protection of the blacks and half-castes' and 'better management of the Aborigines'.

Evangelising the Aboriginals

Intervention in Aboriginal peoples' lives was inextricably linked with bringing them out of 'moral degradation'. Religious humanitarianism played a significant role in the development of Aboriginal affairs in Victoria in the latter half of the nineteenth century, despite the significant impact of a static racial hierarchy on popular racial discourse. However, there was a degree of discord in Christian circles about the fate of Aboriginal people. Some pronounced that 'Australian aborigines were mere beasts in human shape ... and that no efforts made to evangelise the aborigines of Victoria could be successful'. Other prominent

Christians considered that 'the condition of the aborigines is that of dying men' and as all men are created in God's own image, they could be 'saved [from extinction and damnation] only by divine interference'. Aboriginal evangelists in the late 1880s, such as a group of seven men from the Maloga Mission led by Martin Simpson, a Djadjawurrung man, echoed this refrain:

> The idea was to go out among the scattered remnants, preach the gospel and endeavour to gather them at the Mission Station. "We thought a great deal about this", says Martin Simpson, the leader of the little band of Missionaries, "and we looked very long and very earnestly to God that He would open a way for us, and that He might convert us into humble instruments for the salvation of our poor people."

These Aboriginal evangelists also witnessed to the morally degraded non-Indigenous mining community of Ballarat.

> The Aboriginal, Martin Simpson, in addressing a large assemblage at the Alfred hall last night on behalf of the Maloga Mission, stated that since his arrival in the fine City of Ballarat, he felt sorry on finding that there were many white people unconverted to God. During the past few days he was grieved on noticing men staggering about the streets of Ballarat overcome by the effects of intoxicating drink, that firewater which had cut out of existence and sent to a premature grave so many of his (Martin Simpson's) race.

Missionaries, including Daniel, an Aboriginal man from the Lake Hindmarsh region, firmly believed they were acting in the best interests of Aboriginal people, and that to be 'raised' in Christianity was compensation for their losses as a result of British colonisation. Daniel and others such as Fred Wowinda who was observed 'reading the Testament to a black from a neighbouring station', support Richard Broome's contention that 'some Aboriginal people voluntarily embraced cultural enlargement and enrichment'. But not all: in the same period that significant numbers of Aboriginal people chose to express loyalty to 'their Queen', for instance, there were also relatively large numbers near the goldfields who 'distrusted the good intentions of white men' and insistently chose to live independently of the missions and reserves. The *Bendigo Advertiser*, in 1862 reported on the pitiful condition of a Djadjawurrung widow who preferred to stay in her own country rather than be shifted to a foreign reserve or mission. Other Aboriginal voices indicate that Christian sentiments and piecemeal aid were perceived with a degree of ridicule. An Aboriginal-English ditty recorded during the 1850s vividly illustrates Aboriginal peoples' disdain at impoverished charity: 'One blanket 'twee four of us, No jolly good to us, best fellow rum'.

An anecdotal story related by W Dobie describes how 'Prince Jamie', an Aboriginal person at an unspecified goldfield, would provide a scornful performance about the Aboriginal Protection Establishment (Franklinford, central Victoria) to equally scornful gold diggers:

> He [Jamie] would commence with an imitation of the chanting of the service, in a reverent attitude, and with an attempt at grave looks. He would then wind up in solemn accents with these words, "As it was in the beginning, is now, and ever s'all be, world without end, amen. No give it grog, no give it flour, cabbage, baccy, b------- the money!"

In the absence of a pro-active Aboriginal policy by successive Victorian governments, a number of Missionary bodies, most notably the Moravians and Presbyterians, attempted to provide refuge and spiritual teaching. Again, while some Aboriginal people chose to incorporate elements of Christianity into their culture, others embraced Christianity as a means of personal salvation and their race's survival; still others utilised the mission as a refuge and as a platform from which to launch a raft of initiatives which would provide them with greater self-determination. The missionaries were able to impart some precepts of Christianity in the face of an overwhelmingly secular gold frontier society and to provide a nexus for involvement with non-Indigenous people. The lure of the goldfields, the proximity of the goldfields shanties, readily available employment opportunities and a laissez faire attitude towards Aboriginal policy ensured the failure of some missionaries' efforts.

The 1860s heralded an era whereby the colonial government sought to 'protect and control' Aboriginal people's lives. The catalyst for this course of action was pressure brought to bear on the government by Aboriginal and non-Indigenous voices throughout the 1850s to improve the social and economic welfare of Aboriginal people in Victoria. The attempts of the CBA to 'protect and settle' the Victorian Aboriginal population were shown to be largely untenable. Now the very concept of 'protection' was approached from different angles. But again, Aboriginal people responded in a range of ways to the presence of CBA appointed Guardians, particularly in auriferous regions. We have seen how traditional lifestyles were able, in the main, to be continued: George Wathen recorded traditional camping grounds, harming practices, physical ornamentation, housing, clothing, language and earth sculptures being maintained by Djabwurrung people at Challicum Station (central Victoria) in November 1854. Aboriginal voices continued to be raised to entreat for land and other rights; physical refuge was sometimes sought; friendships and kinship ties were forged; new paradigms of living were explored; and the resources of the Guardians were exploited to the advantage of the Aboriginal community. Yet the good intentions of missionaries and the CBA were ultimately shattered by a failure of the colonial governments to commit adequate land and resources, and to implement policy and legislation effectively to fractured Aboriginal communities which they believed were becoming extinct. Adding to the policy discord was the relative absence of consultation with Aboriginal people on matters of policy. The escalation of self-destructive behaviours amongst Aboriginal communities and the increasingly urgent tenor of communication regarding protection from self-abuse floundered in conflicting messages, demands and unreal expectations. Ultimately, what prevailed was confusion as to what to do and a lack of political will to do much at all.

Conclusion

The notion that to the non-Indigenous miners of Victoria Aboriginal people were 'invisible, silent and nameless' has been shown to be false. Vestiges of their considerable physical connection with the goldfields are to be found in Aboriginal artworks of the period, archaeological sites and place names bestowed upon mining areas, as well as the recollections of numerous non-Indigenous miners and contemporary observers. Core motivations for Aboriginal people to engage or not engage in work on the goldfields of Victoria clearly stemmed from whether their ancestral estates rested on auriferous ground and also the kinship-styled relationships that were forged between themselves and the immigrant gold seekers. For some Aboriginal people, just as for non-Indigenous people, gold seeking as a full-time occupation to the exclusion of all other duties was an anathema, and for others it was a worthwhile and productive pursuit.

Mutual interest in each population's 'otherness' cannot be overstated. The Victorian gold miners adoption – or acculturation – of significant Aboriginal cultural features is one of the more intriguing issues in this study that has received very little attention from historians, yet provides a nexus to reconciliation through the process of sharing histories. More research is urgently needed on Aboriginal perspectives of the goldfields. This includes a working knowledge of the dynamics of Aboriginal attachment to land tenure and kinship affiliations in order to explore the variety of sexual, legal, moral, and mercantile arrangements struck.

Moreover, the large body of evidence presented here that documents Aboriginal people actively discovering and seeking gold for their own commercial gain – encompassing immigrating to foreign goldfields, independent fossicking ventures and multi-racial partnerships – has interesting and significant ramifications. Bates' notion of gold as a 'social energiser and definer' can now rightly be applied to Aboriginal society too. For some, gold seeking was consistent to some extent with traditional commercial activity, and, to a point, since many goldfields were probably located where traditional quarries had existed; there was a site usage overlay. Not surprisingly, the records emphasise select language groups (such as the Djadjawurrung) whose successful participation in labouring for gold was prodigious. Many Aboriginal people sought to find their niche in the new society, via predominantly economic channels, through trading in their manufactured goods, farming, and cultural performances, or in employment roles such as bark cutting, tracking, guiding and police work, which did not inordinately compromise their cultural integrity and took advantage of their superior traditional work skills. They also sought to accommodate and manipulate the gold seekers into their social structure by continuing to recognise select non-Indigenous people as resuscitated Aboriginal people and entreating gold miners and townsfolk to adopt an Aboriginal morality. Others chose to align themselves with the goldfields as little as possible. Gold then, spelled a freedom and a social energy rarely enjoyed in the pastoral period, or in subsequent epochs. It is my hope that this work will inspire others to delve deeper into this history – a shared history with great potential for furthering the goal of reconciliation in this country.

Select bibliography

Please note that the PhD thesis on which this book is based contains the full list of references. See: Cahir, David [Fred]. *Black Gold: A History of the Role of Aboriginal people on the Goldfields of Victoria, 1850–70*. PhD thesis, School of Business: University of Ballarat, 2006.

Primary sources

Manuscripts

Annison, G. *Diary*. MS 3878. National Library of Australia (NLA) MS, Canberra.

Anonymous. *Gold Rush Narrative*. MS 6848. NLA MS, Canberra.

Arnot, James. *An Emigrants Journal*. MS 989/3. State Library of Victoria (SLV) MS, Melbourne.

Ballarat Mechanic's Institute Committee. *Committee Minute Book*. Ballarat: Mechanic's Institute, 1879.

Batey, William. *Reminiscences*. MS 035, Box 16. Royal Historical Society of Victoria (RHSV) MS, Melbourne.

Bishop, George. *Memoirs of George Gregor Bishop*. VF 34. RHSV, Melbourne.

Blackman, Nance. *Miscellaneous Papers Re: East Gippsland History*. MS 510. RHSV MS, Melbourne.

Blainey, Samuel. *NilDesperandum*. MS 9398, Box 1123. SLV MS, Melbourne.

Blyth, Thomas. *Diary*. MS 2310. NLA MS, Canberra.

Bond, James. *Memoirs*. Mfm M724. NLA MS, Canberra.

Booth, Thomas. *Diary of Thomas Booth*. Diary. Buninyong Historical Society Archives, Buninyong.

Bridges, Walter. *The Travels of Walter Bridges*. Travelogue. Ballarat.

Buckley, Patrick Coady. *Journal*. MS 000097, Box 37, item 4. RHSV MS, Melbourne.

Burchett, Charles. *Letters*. MS 014506, Box 33–4. RHSV MS, Melbourne.

Chapple, John. *Diary*. MS 11792. SLV MS, Melbourne.

Clutterbuck, Samuel. *Diary*. MS 011230. RHSV MS, Melbourne.

Collyer, Caleb. *Reminiscences*. MS 066, Box 6–7. RHSV MS, Melbourne.

Costello, Patrick. *Patrick Costello; Narrative of His Life as a Port Phillip Pioneer*. MS 776. RHSV MS, Melbourne.

Crown Law Offices. *Inquest Deposition Files*. VPRS 24/P0000. Public Records Office of Victoria (PROV), Melbourne.

Dannock, James. *Autobiography*. Mfm M 1862. NLA MS, Canberra.

Davies, Owen. *Papers*. MS 5552. NLA MS, Canberra.

Eberlie, Charles. *Diary*. MS 034, Box 16/1. RHSV MS, Melbourne.

Fahey, Sen. Const. *Inquest Deposition Files*. VPRS 24, Unit 283, item 1872/232. PROV, Melbourne.

Foreman, Charles. *Letters*. MS 5563. NLA MS, Canberra.

Gilbanks, William. *Letters*. MS 3914. NLA MS, Canberra.

Gow, Robert. *Journal*. MS 24. Australian Institute of Aboriginal and Torres Strait Islanders (AIATSIS) MS, Canberra.

Gray, Henry. *Letters*. MS 6200. NLA MS, Canberra.

Haustorfer, H. *Reminiscences*. MS 7033, Folder 23. NLA MS, Canberra.

Hiscock, GJ. *Descendants of Thomas Hiscock*. Geneological Records. Buninyong.

Horsburgh, Dr. *Correspondence*. VPRS 6605, Unit 10. PROV, Melbourne.

Howitt, AW. *Papers*. MS 9356. SLV MS, Melbourne.

Johns, RE. *Papers*. MSF 10075. SLV MS, Melbourne.

Johnson, Philip. *Papers*. MS 7627. NLA MS, Canberra.

Lagergren, Carl. *Journal and Letters*. MS 3867. NLA MS, Canberra.

Laidler, E. *Maldon*. MS 14618. RHSV MS, Melbourne.

Lazarus, S. *Diary*. MS 11484. SLV MS, Melbourne.

Le Souef, A. *Personal Recollections of Early Victoria*. MS 173. South Australian Museum, Adelaide.

Lousada, Charles. *Old Brandy Creek and Other Reminiscences*. MS 350. RHSV MS, Melbourne.

MacDonald, Sheila. *The Member for Mt Ida*. MS 7033, Folder 17, Box 3. NLA MS, Canberra.

Martin, Thomas. *Early History of Newstead District*. Reminiscences. MS 41074. SLV MS, Melbourne.

McLeish, William. *Memorandum of the Family of John McLeish*. Reminiscences. MS 33169. SLV MS, Melbourne.

Middleton, Edwin. *A Description of the Life and Times in Victoria in the 1860's by a Young Colonist*. Reminiscences. Melbourne.

Morgan, Henry. *Diary*. MS 32281. SLV MS, Melbourne.

Mossman, Martin. *Letters*. MS 33231. SLV MS, Melbourne.

Nawton, William Cussons. *Diary*. MS 41842. SLV MS, Melbourne.

Netell, Edward. *Town Clerks Correspondence*. Letters. Letters From the Town Clerk, Buninyong.

Nisbet, James. *Articles*. MS 3588. NLA MS, Canberra.

O'Grady, Michael. *Inquest Deposition Files*. VPRS 24, Unit 76, item 1859/203. Public Records Office of Victoria, Melbourne.

Panton, J.A. *The Autobiography of J.A Panton*. Autobiography. SLV Manuscripts, Melbourne.

Parker, Edward. *Aborigines: Return to Address*. Melbourne: Victorian Legislative Council, 1854.

Peverell, James. *Reminiscences*. MS 610. RHSV MS, Melbourne.

Pierson, Thomas. *Diaries*. MS 11646. SLV MS, Melbourne.

Pope, Richard. *Diaries*. Diary. MS 11918. SLV MS, Melbourne.

Price, Edwin. *Letters*. MS 4826. NLA MS, Canberra.

PROV. *Day Book of the Native Police Corps, Narre Warren*. VPRS 90, Unit 1. VPRS, Melbourne.

—. *Letter from Commissioner of Crown Lands to La Trobe*. Series 103, Unit 1, item 48/26. VPRS, Melbourne.

Radden, W. *The Early History of Warrandyte 1839–66*. MS 324. RHSV, Melbourne.

Rayment, Raymond. *Diary*. M4471. SLV MS, Melbourne.

Rodaughan, John. *Brief History of Ringwood East and Burnt Bridge*. MS 384. SLV MS, Melbourne.

Rogers, Dorothy. *A Kew Centenarian*. Typescript of newspaper article with notes. MS 46759. Morgan, James, SLV MS, Melbourne.

Rowe, George. *Correspondence*. Letters. MS 3116. NLA MS, Canberra.

Rutherford, James. *Central Board for the Protection of the Aborigines*. Correspondence Files, B313, item 41. National Australian Archives, Melbourne.

Salisbury, Robert. *Letter*. MS 210. NLA MS, Canberra.

Selby, James. *Diary and Papers*. MS 9866. SLV MS, Melbourne.

Shellard, Frank. *Reminiscences of an Old Digger*. MS 1890. NLA MS, Canberra.

Sinclair, James. *Memoirs*. MS 5052. NLA MS, Canberra.

Smyth, Robert Brough. *Correspondence*. MS 1176/5. SLV MS, Melbourne.

Stanthorpe, Robert. *Reminiscences*. MS 7033. NLA MS, Canberra.

Strutt, William. *Off to Australia*. MS 15886. NLA MS, Canberra.

Sugden, George. *Pioneering Life in Outback Stations of Victoria*. MS 301. RHSV MS, Melbourne.

Tame, Robert. *Reminiscences of Melbourne and Gold Diggings*. Reminiscences. MS 8964. SLV, Melbourne.

Thomas, Robert. *Autobiography*. Mfm M2090. NLA MS, Canberra.

Thomas, William. *Sketch of Manners*. Thomas Papers. SLV MS, Melbourne.

—. *Journal*. Ms 214. Mitchell Library, Sydney.

—. *Registered Inward Correspondence to the Surveyor General from the Guardian of Aborigines*. VPRS 2894. PROV, Melbourne.

Thomson, JC. *Correspondence*. VPRS 103, Unit 1, item 1. PROV, Melbourne.

Tomlinson, William. *Diary*. MS 12183. SLV, Melbourne.

Unknown. *Evansford History*. MS 7033, Folder 8. NLA MS, Canberra.

—. *Blacks on Barham Station (Denbeigh) Murray River*. MS 7033, Folder 14, Box 2. NLA MS, Canberra.

—. *Colonial Secretary Inward Correspondence*. VPRS 2878, unit 1, item 51/417. PROV, Melbourne.

—. *Colonial Secretary's Office: Correspondence*. VPRS 3219, Volume 1. PROV, Melbourne.

—. *Colonial Secretary's Office: Correspondence*. VPRS 1189, unit 16, folder 6, items 2588, 4831, 2783, 3100, 3589, and 3751. PROV, Melbourne.

Wakefield, George. *Letters*. MS 684. NLA MS, Canberra.

Walker, JH. *Memoirs of Joseph Henry Walker*. MS 11877. SLV MS, Melbourne.

White, Emma. *Letters*. MS 3878. NLA MS, Canberra.

Whittle, Robert. *Reminiscences*. MS 11829, Box 2163/7. SLV MS, Melbourne.

Woodbury, Walter. *Letters*. Mfm M 1952. NLA MS, Canberra.

Woolner, Thomas. *Diary*. MS 2939. NLA MS, Canberra.

Published sources

Anonymous. *Pentland Hills*. Illustrated Australian Magazine. Vol. 3–4. Melbourne: Ham Brothers, 1852.

Blandowski, W. *Personal Observations in the Central Parts of Victoria*. Melbourne: Goodhugh and Trembath, 1855.

Board for the Protection of Aborigines. *Report[s] of the Board for the Protection of Aborigines in the Colony of Victoria. Presented to Both Houses of Parliament*. Melbourne: BPA, 1871–1895.

Bonwick, James. *Notes of a Gold Digger and Gold Digger's Guide*. Melbourne: Connebee, 1852.

Bride, TF, ed. *Letters from Victorian Pioneers*. Melbourne: William Heinemann, 1898.

Brout, C. *Guide for Emigrants to the Australian Gold Mines*. Trans. Didier Leclere. Paris: Unknown, 1861.

Butler, Samuel. *The Gold Regions of Australia. Who Ought to Go to the Diggings and Who Ought to Remain at Home*. Glasgow: McPhun, 1858.

Caldwell, Robert. *The Gold Era of Victoria*. Melbourne: Blundell, 1855.

Campbell, A. *Rough and Smooth or Ho! For an Australian Goldfield*. Quebec: Hunter and Rose, 1865.

Carboni, Raffaelo. *Gilburnia*. 1872. Trans. Tony Paglario. Melbourne: Jim Crow Press, 1993.

Castella, Hubert de. *Australian Squatters*. 1861. Trans. CB Thornton-Smith. Melbourne: MUP, 1987.

Castlemaine Association of Pioneers and Old Residents. *Records of the Castlemaine Pioneers*. Melbourne: Rigby, 1972.

Central Board of Aborigines. *Report of the Central Board Appointed to Watch over the Interests of the Aborigines in the Colony of Victoria. Presented to Both Houses of Parliament*. Melbourne: Victoria Parliament, 1861.

—. *Report of the Central Board Appointed to Watch over the Interests of the Aborigines in the Colony of Victoria*. Melbourne: Victoria Parliament, 1862.

—. *Third Report of the Central Board Appointed to Watch over the Interests of the Aborigines in the Colony of Victoria*. Melbourne: Victoria Parliament, 1863.

—. *Fourth Report of the Central Board Appointed to Watch over the Interests of the Aborigines in the Colony of Victoria*. Melbourne: Victoria Parliament, 1864.

—. *Fifth Report of the Central Board Appointed to Watch over the Interests of the Aborigines in Victoria*. Melbourne: Victoria Parliament, 1866.

—. *Sixth Report of the Central Board Appointed to Watch over the Interests of the Aborigines in the Colony of Victoria*. Melbourne: Victoria Parliament, 1869.

Chabrillan, Celeste de. *The French Consul's Wife: Memoirs of Celeste De Chabrillan in Goldrush Australia*. 1877. Trans. P Clancy and J Allen. Melbourne: Miegunyah Press, 1998.

Chistiakov, M. *Tales from a Journey through Australia*. St Petersburg: Unknown, 1874.

Clacy, Charles, Mrs. *A Lady's Visit to the Gold Diggings of Australia, in 1852–3: Written on the Spot*. London: Hurst and Blackett, 1853.

Clacy, D. *Lights and Shadows of Australian Life*. London: Hurst and Blackett, 1854.

Comettant, Oscar. *In the Land of Kangaroos and Gold Mines*. 1890. Trans. Judith Armstrong. Melbourne: Rigby, 1980.

Coningsby, Robert. *The Discovery of Gold*. London: William Milligan, 1895.

Craig, William. *My Adventures on the Australian Goldfields*. London: Unknown, 1903.

Curr, EM. *The Australian Race*. Melbourne: John Ferres, 1886.

Dunderdale, George. *The Book of the Bush*. 1870. Ringwood: Penguin Books Ltd, 1973.

Erskine, John. *A Short Account of the Late Discoveries of Gold in Australia*. London: Boone, 1852.

Evans, GC. *Stories Told around the Camp Fire: Compiled from the Notebook of D. Digwell*. Bendigo: Bendigo Independent Office, 1881.

Ewes, JD. *China, Australia and the Pacific Islands, in the Years 1853–56*. London: Richard Bentley, 1857.

Fauchery, Antoine. *Letters from a Miner in Australia*. 1857. Trans. A.R Chisholm. Melbourne: Georgian House, 1965.

Ferguson, C. *Experiences of a Forty-Niner in Australia and New Zealand*. 1888. Melbourne: Gaston Renard, 1979.

Foster, JF. *The New Colony of Victoria Formerly Port Phillip...And Introducing All the Latest Information to December 1851*. London: Trelawney Saunders, 1851.

Graham, John, ed. *Observations and Experiences During 25 Years of Bush Life in Australia*. London: Book Society, 1863.

Hall, William. *Practical Experience at the Diggings of the Gold Fields*. London: Effingham, 1852.

Hamilton, JC. *Pioneering Days in Western Victoria*. Kowree: Shire of Kowree, ca 1912.

Hancock, Marguerite, ed. *Glimpses of Life in Victoria by a Resident*. 1876. Melbourne: MUP, 1996.

Haydon, A. *The Trooper Police of Australia*. London: Melrose, 1911.

Howitt, AW. *The Native Tribes of South East Australia*. Melbourne: Macmillan, 1904.

Howitt, William. *Land, Labour and Gold*. 1855. 2nd ed. Kilmore: Lowden, 1972.

Jones, Henry. *Adventures in Australia in 1852 and 1853*. London: Bentley, 1853.

Journalist, Australian. *The Emigrant in Australia*. London: Addey, 1852.

Journet, F. *L'australie: Description Du Pays*. Trans. Etienne Lambert. Paris: Rothschild, 1885.

Just, P. *Australia: Or, Notes Taken During a Residence in the Colonies from the Gold Discovery in 1851 Till 1857*. Dundee: Durham and Thomson, 1859.

Kirkland, Katherine. *Life in the Bush by a Lady*. Edinburgh: Chambers, 1845.

Lancelott, F. *Australia as It Is: It's Settlements, Farms and Goldfields*. 1852. Tokyo: Charles Tuttle, 1967.

Lang, John. *The Australian Emigrants Manual*. London: Partridge, 1852.

Mackay, George. *Annals of Bendigo, 1851 to 1867*. Bendigo: Mackay, 1912.

—. *History of Bendigo*. Bendigo: Lerk and McClure, 2000.

Mackay, Richard. *Recollectionsof Early Gippsland Goldfields*. Traralgon: W. Chappell, 1916.

McCombie, Thomas. *Australian Sketches*. London: W. Johnson, 1861.

McK, Katherine. *Old Days and Gold Days*. Melbourne: McK, 1910.

Meredith, Louisa. *Over the Straits: A Visit to Victoria*. London: Chapman and Hall, 1861.

Mereweather, JD. *Diary of a Working Clergyman in Australia, 1850–3*. London: Unknown, 1859.

Mossman, S, ed. *Emigrants' Letters from Australia*. London: Addey and Co., 1853.

Mossman, Samuel. *The Gold Regions of Australia*. London: William Orr and Co, 1852.

Mossman, Samuel, and Thomas Bannister. *Australia, Visited and Revisited*. 1853. Sydney: Ure Smith, 1974.

New South Wales Legislative Council. *Report from the Select Committee on the Condition of the Aborigines, with Appendix, Minutes of Evidence and Replies to a Circular Letter*. New South Wales Legislative Council, Sydney, 1849.

Parker, Edward. *The Aborigines of Australia: A Lecture*. Melbourne: Hugh McColl, 1854.

Patterson, JA. *The Gold Fields of Victoria in 1862*. Melbourne: Robertson, 1862.

Polehampton, Reverend Arthur. *Kangaroo Land*. London: Richard Bentley, 1862.

Prout, JS. *An Illustrated Handbook of the Voyage to Australia*. London: Peter Duff, 1852.

Ramsay-Laye, Elizabeth. *Social Life and Manners in Australia*. London: Longman and Green, 1861.

Read, Rudston. *What I Heard, Saw and Did at the Australian Goldfields*. London: Boone, 1853.

Religious Tract Society. *Australia and Its Settlements*. London: Religious Tract Society, 1853.

—. *Australia: Its Scenery, Natural History, Resources and Settlements with a Glance at Its Goldfields*. London: Religious Tract Society, ca 1854.

Roberts, Morley. *King Billy of Ballarat: And Other Short Stories*. London: Lawrence and Bullen, 1892.

Robertson, William. *History of Piggoreet and Golden Lake*. 1927. Facsimile ed. Daylesford: Jim Crow Press, 1998.

Rochford, John. *The Adventures of a Surveyor in New Zealand and Australia*. London: David Bogue, 1853.

Sadlier, John. *Recollections of a Victorian Police Officer*. London: Penguin, 1913.

Sherer, John. *The Gold-Finder of Australia*. 1853. Facsimile ed. Ringwood: Penguin, 1973.

Sidney, Samuel. *The Three Colonies of Australia: NSW, Victoria and South Australia*. London: Ingram and Cooke, 1852.

Smyth, Robert Brough. *The Aborigines of Victoria*. Melbourne: Government Printer, 1878.

Sutherland, A. *Victoria and Its Metropolis: Past and Present*. Melbourne: Macarron Bird, 1888.

Sutherland, George. *Tales of the Goldfields*. Melbourne: George Robertson, 1880.

Therry, Roger. *Reminiscences of Thirty Years Residence in New South Wales*. Facsimile ed. Sydney: SUP, 1974.

Tucker, J. *Mission to the Chinese and Aborigines*. Sydney: Joseph Cook and Co, 1868.

Unknown. *Ballarat*. Illustrated Australian Magazine. Vol. 3–4. Melbourne: Ham Brothers, 1852.

Victorian Government. *Report of the Select Committee of the Legislative Council on the Aborigines; Together with the Proceedings of Committee, Minutes of Evidence, and Appendices*. Melbourne: John Ferres, Government Printer, 1859.

Victorian Royal Commission. *Royal Commission on the Aborigines*. Melbourne: Government Printer, 1877.

Wathen, George. *The Golden Colony, or Victoria in 1854: With Remarks on the Geology of the Australian Gold Fields*. London: Longman, 1855.

Westgarth, William. *Victoria and the Australian Gold Mines in 1857*. London: Smith, Elder and Co, 1857.

Westwood, JJ. *Journal of JJ Westwood Being an Account of Eight Years Itinerary to the Townships and Squatting Stations of Victoria*. Melbourne: Clarson, Shallard & Co, 1865.

Wheelwright, H. *Bush Wanderings of a Naturalist*. 1861. Melbourne: Oxford University Press, 1979.

Wilmer, K. *Adventures at the Goldfields*. London: Dean and Son, 1859.

Newspapers

Albury Post

Ararat Advertiser

Argus

Ballarat Star

Ballarat Times

Border Post

Corowa Free Press

Geelong Advertiser

Illustrated Melbourne Post

London Times

Mount Alexander Mail

Mount Ararat and Pleasant Creek Advertiser

Ovens and Murray Advertiser

Riponshire Advocate

Secondary sources

Adamthwaite, Vera. *The Story of the Family Called White; 1852–1982*. Ballarat: Waller and Chester, 1982.

Anderson, Hugh. *The Flowers of the Field – a History of Ripon Shire*. Melbourne: Hill of Content, 1969.

Annear, R. *Nothing but Gold: The Diggers of 1852*. Melbourne: Text Publishing, 1999.

Attwood, Bain. *My Country – a History of the Djadja Wurrung; 1837–1864*. Clayton: Monash Publications, 1999.

Bancroft, Robyne. 'Aboriginal Miners and the Solferino and Lionsville Goldfields of Northern New South Wales'. *A World Turned Upside Down: Cultural Change on Australia's Goldfields 1851 – 2001*. Eds. Kerry Cardell and Cliff Cumming. Canberra: ANU, 2001: 131–147.

Banfield, Lorna. *Like the Ark: The Story of Ararat*. Melbourne: Cheshire, 1956.

Barrett, Charles. *Gold: The Romance of Its Discovery in Australia*. Melbourne: United Press, 1944.

Barwick, Laura, and R Barwick, eds. *Rebellion at Coranderrk*. Monograph 5. Canberra: Aboriginal History, 1998.

Bate, Weston. *Lucky City*. Melbourne: Melbourne University Press (MUP), 1978.

Beavis, Margery. *Avoca – the Early Years*. Warrnambool: Margery and Betty Beavis, 1986.

Blainey, Geoffrey. *The Rush That Never Ended*. 1963. 5th ed. Melbourne: MUP, 2003.

Blake, Les, ed. *A Gold Digger's Diaries by Ned Peters*. Newtown: Neptune Press, 1981.

Blake, L. *Gold Escorts in Australia*. Melbourne: Macmillan, 1993.

Blanks, Harvey. *The Story of Yea*. Melbourne: Hawthorn Press, 1973.

Bridges-Webb, C. *The Goldfields of Gippsland*. Traralgon: Traralgon and District Historical Society, 1969.

Broome, Richard. *Aboriginal Victorians: A History since 1800*. Sydney: Allen and Unwin, 2005.

—. 'Professional Aboriginal Boxers in Eastern Australia 1930–1979'. *Aboriginal History* 4.1 (1980): 49–73.

—. 'Aboriginal Workers on South-Eastern Frontiers'. *Australian Historical Studies* 26.103 (1994): 202–220.

Brown, PL, ed. *Clyde Company Papers Volumes 1–5*. London: Oxford University Press (OUP), 1971.

Cadzow, G, and N Wright. *Charlton in the Vale of the Avoca*. Charlton: Charlton Bi-Centenary Committee, 1988.

Cannon, M, ed. *Aborigines of Port Phillip 1835–1839*. Melbourne: VGPO, 1982.

Carmody, Jean. *Early Days of the Upper Murray*. Wangaratta: Shoestring Press, 1981.

Carrodus, Geraldine. *Gold, Gamblers and Sly Grog: Life on the Goldfields, 1851–1900*. Melbourne: OUP, 1981.

Christie, Michael. *Aboriginals in Colonial Victoria*. Melbourne: MUP, 1979.

Clark, Ian. *Aboriginal Languages and Clans: An Historical Atlas of Western and Central Victoria*. Melbourne: Monash University, 1990.

—. *'That's My Country Belonging to Me'. Aboriginal Land Tenure and Dispossession in Nineteenth Century Western Victoria*. Ballarat: Ballarat Heritage Services, 2003.

Clark, Ian, and David Cahir. 'Aboriginal People, Gold and Tourism: The Benefits of Inclusiveness for Goldfields Tourism in Regional Victoria'. *Tourism, Culture and Communication* 4 (2003): 1–14.

Collett, Barry. *Wednesdays Closest to the Full Moon*. Carlton: MUP, 1994.

Colquhuon, S. *Mitta Mitta from the Early Pioneer Days*. Colquhoun, 1953.

Coxall, Douglas, ed. *Coxall Family History*. Buninyong: Pearl Winn, 1992.

Critchett, Jan. *A Distant Field of Murder; Western District Frontiers 1834–1888*. Melbourne: MUP, 1990.

Cuffley, Peter, ed. *Send the Boy to Sea: The Memoirs of a Sailor on the Goldfields by James Montagu Smith*. Noble Park: Five Mile Press, 2001.

Cusack, Frank, ed. *Early Days on Bendigo*. Melbourne: Queensberry Hill Press, 1979.

—, ed. *The History of the Wedderburn Goldfields*. 1888. Carlton: Queensberry Hill Press, 1981.

Daley, Charles. *The Story of Gippsland*. Melbourne: Whitcombe & Tombs, 1962.

Donaldson, T, ed. *This Is What Happened; Historical Narratives by Aborigines*. Canberra: Australian Institute of Aboriginal Studies (AIAS), 1986.

Duyker, Edward, ed. *A Woman on the Goldfields: Recollections of Emily Skinner, 1854–1878*. Carlton: MUP, 1995.

Evans, William, ed. *Diary of a Welsh Swagman, 1869–1898*. South Melbourne: Macmillan, 1977.

Fabian, Susan, ed. *Mr Punch Down Under*. Melbourne: Fabian Green House, 1982.

Fead, George. 'Notes of an Unsettled Life'. *Gippsland Heritage Journal* 16 (1994): 24–36.

Fels, Marie. *Good Men and True: The Aboriginal Police of the Port Phillip District, 1837–1853*. Melbourne: MUP, 1988.

Finlay, Alexander. *The Journal of Alexander Finlay at the Victorian Gold Diggings*. Sydney: St Marks Press, 1992.

Fletcher, BJ, ed. *Broadford: A Regional History*. Melbourne: Lowden, 1975.

Flett, James. *Dunolly: Story of an Old Gold Diggings*. Melbourne: Hawthorn Press, 1974.

—. *Maryborough*. Blackburn: Dominion Press, 1975.

Foxcroft, EJ. *Australian Native Policy: Its History, Especially in Victoria*. Melbourne: MUP, 1941.

Goodman, David. *Gold Seeking: Victoria and California in the 1850s*. St Leonards: Allen and Unwin, 1994.

—. 'Making an Edgier History of Gold'. *Gold: Forgotten Histories and Lost Objects of Australia*. Ed. Iain McCalman. Cambridge: CUP, 2001: 23–37.

Griffiths, Tom, ed. *The Life and Adventures of Edward Snell*. North Ryde: Angus and Robertson, 1988.

Halls Gap & Grampians Historical Society. *Victoria's Wonderland*. Halls Gap: Halls Gap & Grampians Historical Society, 2006.

Hibbins, G. *Barmah Chronicles*. Lynedoch: Lynedoch Publications, 1991.

Hudson, Mike. *Bound for the Goldfields: A True Account of a Journey from Melbourne to Castlemaine*. Katoomba: Wayzgoose Press, 1990.

James, GF, ed. *A Homestead History – Being the Reminiscences and Letters of Alfred Joyce of Plaistow and Norwood, Port Phillip 1843 to 1864*. Melbourne: OUP, 1969.

Keesing, Nancy, ed. *History of the Australian Gold Rushes*. Sydney: Angus and Robertson, 1971.

Kellerman, M. 'Interesting Account of the Travels of Abraham Abrahamsohn'. *Australian Jewish Historical Society Journal* 7 (1974): 478–494.

Krefft, Gerard. 'On the Manner and Customs of the Aborigines of the Lower Murray and Darling'. *Journal and Proceedings of the Philosophical Society of New South Wales* (1862–65): 359–371.

Leatherbee, Mrs Albert, ed. *Knocking About: Being Some Adventures of Augustus Baker Pierce in Australia*. Wangaratta: Shoe String Press, 1984.

Leslie, John, and H Cowie, eds. *The Wind Still Blows: Early Gippsland Diaries*. Sale: HC Cowie, 1973.

Little, William. *William Little of Ballarat: Some Writings. 1887–1907*. Ed. Frederick Shade. Mitcham: Eastside Printing, 2001.

Longmire, Anne. *Nine Creeks to Albacutya – a History of the Shire of Dimboola*. North Melbourne: Hargreen Publishing Co, 1985.

Mackaness, George, ed. *Murray's Guide to the Gold Diggings: The Australian Gold Diggings*. Vol. 8. Sydney: D.S. Ford, 1956.

Mallacoota and District Historical Society. *Mallacoota Memories*. Mallacoota Historical Society, 1980.

McBryde, Isabel. 'Exchange in South Eastern Australia: An Ethnohistorical Perspective'. *Aboriginal History* 8.2 (1984): 132–153.

McCalman, Iain, ed. *Gold: Forgotten Histories and Lost Objects of Australia.* Cambridge: Cambridge University Press (CUP), 2001.

McDonald, Barry. 'Evidence of Four New England Corroboree Songs Indicating Aboriginal Responses to European Invasion'. *Aboriginal History* 20.1 (1996): 176–195.

McGivern, Muriel. *Big Camp Wahgunyah: History of the Rutherglen District.* Melbourne: Spectrum, 1983.

Monaghan, Jay. *Australians and the Gold Rush.* Los Angeles: University of California Press, 1966.

Moore, Laurie. *Shot for Gold; the Murder of Thomas Ulrick Burke on the Woady Yalloak Goldfield.* Daylesford: Jim Crow Press, 2002.

Palmer, Joan Austin, ed. *William Moodie: A Pioneer of Western Victoria.* Maryborough: Hedges & Bell Pty Ltd, ca 1975.

Parsons, Michael. 'The Tourist Corroboree in South Australia'. *Aboriginal History* 21.1 (1997): 46–69.

Pearson, A. *Echoes from the Mountains.* Omeo: Omeo Shire, 1969.

Presland, Gary. *For God's Sake, Send the Trackers.* Melbourne: Victoria Press, 1998.

Ragless, Margaret, ed. *Oliver's Diary: An 'Andkerchief of Eirth.* Hawthorndene: Investigator Press, 1986.

Randell, John. *Kimbolton.* Melbourne: Queensberry Press, 1976.

—. *Shire of Mcivor.* Burwood: Anderson, 1985.

Reynolds, Henry. *The Other Side of the Frontier.* Melbourne: Penguin, 1982.

—. *With the White People: The Crucial Role of Aborigines in the Exploration and Development of Australia.* Ringwood: Penguin, 1990.

—. *Black Pioneers.* Ringwood: Penguin, 2000.

Rich, Margaret, ed. *Eugene Von Guerard in Ballarat: Journal of an Australian Gold Digger by Eugene Von Guerard.* Ballarat: Ballarat Fine Art Gallery, 1990.

Robe, Stanley, ed. *Seweryn Korzelinski: Memoirs of Gold-Digging in Australia.* St Lucia: University of Queensland Press (UQP), 1979.

Ronan, M, ed. *Early Dederang 1854–1956 from the Notebook of Michael James Goonan.* Melbourne: Macron, 2004.

Salisbury, Robert. *Lord Robert Cecil's Gold Fields Diary*. Melbourne: MUP, 1935.

Sayers, CE, ed. *Western Victoria – the Narrative of an Educational Tour in 1857*. Melbourne: Heinemann, 1970.

Shelmerdine, Stephen. 'The Port Phillip Native Police Corps'. Hons thesis. University of Melbourne, 1972.

Stanbridge, WE. 'Some Particulars of the General Characteristics of the Astronomy and Mythology of the Tribes in the Central Parts of Victoria'. *Transactions of the Ethnological Society of London* 1 (1861): 286–304.

Stone, Caitlin. 'Researching Native Police Records at PROV'. *Proactive* (2005): 10–11.

Urry, James, and Carol Cooper. 'Art, Aborigines and Chinese: A Nineteenth Century Drawing by the Kwat Kwat Artist Tommy Mcrae'. *Aboriginal History* 5.1–2 (1981): 81–88.

Walker, Scott McLeod. *Glenlyon Connections*. Stanthorpe: Walker, 1993.

Wallace, Ray. *Eaglehawk: Sketch Book of a Golden Past*. Bendigo: Cambridge Press, 1983.

Wesson, Sue. 'The Aborigines of Eastern Victoria and Far South-Eastern New South Wales, 1830–1910: An Historical Geography'. PhD thesis. Monash University, 2002.

Wilkinson, Thomas. 'A Record of Olden Days'. *Journal of Wagga Wagga and District Historical Society* 3 (1970): 3–13.

Williams, AJ. *A Concise History of Maldon and the Tarrangower Diggings*. 1953. Maldon: Tarrangower Times, 1987.

Index